MAXIMUM PRINCIPLES AND THEIR APPLICATIONS

This is Volume 157 in
MATHEMATICS IN SCIENCE AND ENGINEERING
A Series of Monographs and Textbooks
Edited by RICHARD BELLMAN, *University of Southern California*

The complete listing of books in this series is available from the Publisher upon request.

MAXIMUM PRINCIPLES AND THEIR APPLICATIONS

René P. Sperb

Seminar für Angewandte Mathematik
Zurich, Switzerland

ACADEMIC PRESS

A Subsidiary of Harcourt Brace Jovanovich, Publishers

New York London Toronto Sydney San Francisco 1981

ACADEMIC PRESS, INC.
111 Fifth Avenue, New York, New York 10003

United Kingdom Edition published by
ACADEMIC PRESS, INC. (LONDON) LTD.
24/28 Oval Road, London NW1 7DX

Library of Congress Cataloging in Publication Data

Sperb, René Peter, Date
Maximum principles and their applications.

(Mathematics in science and engineering;)
Bibliography: p.
Includes index.
1. Differential equations, Partial--Numerical solutions. 2. Maximum principles (Mathematics)
I. Title. II. Series.
QA377.S68 515.3'53 81-2436
ISBN 0-12-656880-4

PRINTED IN THE UNITED STATES OF AMERICA

81 82 83 84 9 8 7 6 5 4 3 2 1

CONTENTS

PREFACE

A classical result in partial differential equations states that if h is a harmonic function in some bounded domain D of Euclidean space, then the quantity $|\nabla h|^2$ (∇: gradient) takes its maximum on the boundary ∂D. A closely related result is true for a biharmonic function b: It was shown by Miranda (1948) that if D is a plane domain, then the quantity $|\nabla b|^2 - b\,\Delta b$ (Δ: Laplacian) assumes its maximum on ∂D. An important special case of a biharmonic function is the torsion function ψ, which is the solution of

$$\Delta\psi + 1 = 0 \quad \text{in} \quad D, \qquad \psi = 0 \quad \text{on} \quad \partial D.$$

For ψ, Miranda's result thus states that the function

$$P_1 := |\nabla\psi|^2 + \psi$$

assumes its maximum on ∂D. Motivated by Miranda's result, Payne (1968) showed that for the torsion function ψ, the associated function

$$P_2 := |\nabla\psi|^2 + 2\psi$$

attains its maximum where $\nabla\psi = 0$, i.e., where ψ is a maximum, provided D is a convex domain. In the one-dimensional case, the torsion function is a solution of

$$\frac{d^2\psi}{dx^2} + 1 = 0$$

in some finite interval (a, b) of the real line. One easily verifies then that the corresponding quantity P_2 defined above, i.e.,

$$\left(\frac{d\psi}{dx}\right)^2 + 2\psi,$$

reduces to a constant. If we now consider a solution u of

$$\frac{d^2u}{dx^2} + f(u) = 0 \qquad \text{in} \quad (a, b),$$

it is easily checked that the quantity

$$P := \left(\frac{du}{dx}\right)^2 + 2\int_0^u f(y)\,dy$$

is just a constant. This in turn suggests considering the analogous expression

$$P := |\nabla u|^2 + 2\int_0^u f(y)\,dy$$

if u is a solution of

$$\Delta u + f(u) = 0 \qquad \text{in} \quad D, \qquad\qquad u = 0 \qquad \text{on} \quad \partial D.$$

Indeed, it turns out again that the function P attains its maximum at a point where $\nabla u = 0$ if ∂D is convex. This result was first found by Payne and Stakgold (1972), and it marked the beginning of a series of papers that were all concerned with various generalizations and applications of maximum principles for such a function P associated with the solution of some boundary value problem. These maximum principles proved to be a very useful tool in deriving all kinds of a priori bounds. In addition, most of the bounds found by these methods have the nice feature of being optimal in some sense. This means that the inequalities become equalities in certain nontrivial cases.

This volume gives a presentation of the research of the past ten years on such applications of maximum principles in elliptic and parabolic problems. To a large extent this book is self-contained, and only basic knowledge of partial differential equations is required.

The first chapter contains a short description of some problems from different fields leading to elliptic and parabolic equations. In addition, some indications are given about the role of a priori inequalities.

A brief presentation of the main tool, namely, the maximum principles in elliptic and parabolic problems, follows in Chapter 2. In Chapter 3 the method of sub- and supersolutions is used in order to prove existence of solutions in elliptic and parabolic problems. In addition, there is a short survey on regularity results. The last preparatory section, Chapter 4, reviews some differential geometry and tensor analysis. Chapter 5 could probably be called the core of this book, introducing the key quantity, the so-called P-function described in previous chapters for the simplest cases. First applications in some problems of physical interest follow in Chapter 6. The next four chapters deal with various generalizations. In Chapter 7 elliptic equations in divergence form are considered, and in Chapter 8 the Laplacian is replaced by an elliptic operator. In Chapter 9 some generalizations to parabolic problems are given, and still further extensions are mentioned in Chapter 10. These extensions include fourth-order problems and systems. In the last chapter, a number of results are proven that are related in spirit to the previous chapters. Included are the use of level-surface coordinates for the derivation of various isoperimetric inequalities as well as some results on maximum principles involving harmonic functions.

It is hoped that this book may stimulate researchers to apply the results cited herein, or to develop techniques similar to the ones presented here, thereby closing some gaps that are still left open.

It is a pleasure to thank Prof. L. E. Payne of Cornell University for the many stimulating discussions I had with him on the subject of this book, for which his research has laid the cornerstone.

Chapter 1
INTRODUCTION

1.1 NOTATION AND SOME BASIC DEFINITIONS

Let D be a finite domain of n-dimensional Euclidean space E^n. The boundary of D will be designated by ∂D and we write $\bar{D}$ for the closed domain $D \cup \partial D$.

Let $u(x)$ be a real function defined for every point $x = (x^1, x^2, \ldots, x^n)$ of D. We then denote partial derivatives $\partial u/\partial x^i$ by $u_{,i}$ and use the summation convention, so that, e.g., the square of the gradient of u may be written as

$$|\nabla u|^2 = \sum_{i=1}^{n} \left(\frac{\partial u}{\partial x^i}\right)^2 =: u_{,i}u_{,i}.$$

The scalar product of the gradients of two functions, say u and v, will be written as

$$\nabla u \cdot \nabla v = u_{,i}v_{,i},$$

and the Laplacian becomes

$$\Delta u = u_{,ii}.$$

The distance between two points x, y of E^n will be denoted by $|x - y|$, the volume element of D by dx, and the surface element of ∂D will usually be written as ds.

A function $u(x)$ is said to be "Hölder continuous with exponent α" or just "α-Hölder continuous" in D if the expression

$$|u(x) - u(y)|/|x - y|^\alpha$$

is bounded from above for $x \neq y$ (x and y points of D). Here $0 < \alpha \leq 1$, and for $\alpha = 1$, the function u is called Lipschitz continuous. The number

$$\sup_{x, y \in D} \frac{|u(x) - u(y)|}{|x - y|^\alpha}$$

is usually called the minimum Hölder coefficient of u. A function $u(x)$ defined in D possessing k α-Hölder continuous derivatives (i.e., all possible kth derivatives of u) in D is said to be of class $C^{k+\alpha}$ in D, or short: $u \in C^{k+\alpha}(D)$.

Assume that any portion Γ of the boundary of ∂D (an $(n-1)$-dimensional hypersurface in E^n) can be represented in parametric form as

$$\Gamma: \quad x^i = x^i(\xi^\beta), \qquad \beta = 1, \ldots, n-1.$$

Then, if all the functions $x^i(\xi^\beta)$ and their derivatives up to order k are α-Hölder continuous in their respective domains of definition, ∂D itself is said to be of class $C^{k+\alpha}$, or short: ∂D is $C^{k+\alpha}$.

Of course it has been assumed that only admissible representations have been used, i.e., the function $x^i(\xi^\beta)$ satisfy

$$x^i{}_{,\beta} x^i{}_{,\beta} > 0.$$

On ∂D, u can be considered as a function of the ξ^β on any portion Γ of ∂D. Then if $u(\xi^\beta) = u(x^i(\xi^\beta))$ turns out to be of class $C^{k+\alpha}(\Gamma)$ for any portion Γ of ∂D, which is itself of class $C^{p+\alpha}$, $p \geq k$, we shall just say that u is of class $C^{k+\alpha}$ on ∂D or $u \in C^{k+\alpha}(\partial D)$.

The notations $C^{k+\alpha}(D)$ and $C^{k+\alpha}(\partial D)$ will be of importance in Chapter 3 in particular. For most parts of this book it is always tacitly assumed that all functions have as many continuous derivatives as are used in the calculations. If necessary, the smoothness properties required will be stated again explicitly.

A "classical" solution of a problem will be a solution having at least as many continuous derivatives as appear in the corresponding differential equations and boundary conditions. We shall only be interested in classical solutions, and hence "solution" always means "classical solution." For example, by a solution of the Poisson problem

$$\Delta u + \rho(x) = 0 \quad \text{in} \quad D, \qquad u = 0 \quad \text{on} \quad \partial D,$$

where ρ is a given continuous function in D, we mean a function of class $C^2(D) \cap C(\partial D)$. Here $C(\partial D) \equiv C^0(\partial D)$ denotes, of course, the class of continuous functions on ∂D.

1.2 SOME EXAMPLES OF PROBLEMS LEADING TO ELLIPTIC EQUATIONS

1.2.1 Torsion of an Elastic Beam

Let D be a simply connected cross section of a cylindrical bar that is twisted by terminal couples. If the angle of twist per unit length is sufficiently small, one is led to the so-called Sain Venant torsion problem. It can be formulated mathematically as follows: we seek a solution $u(x)$ of

$$\Delta u = -2 \qquad \text{in} \quad D \subset E^2, \tag{1.1}$$

vanishing on the boundary ∂D. The components of the resulting stress are then given by $\mu\alpha(u_{,2}, -u_{,1})$ where μ is the shear modulus and α the angle of twist per unit length. The magnitude of the shearing stress is thus given by

$$\tau = \mu\alpha|\nabla u|. \tag{1.2}$$

The torsion problem is a model problem for some of the methods to be presented in this book and will be discussed in detail.

With this mechanical interpretation of Eq. (1.1), we are mainly interested in obtaining information about τ. For example, where does the maximum stress occur and what bounds in terms of the geometry of ∂D can one give? Another quantity of physical interest is the torsional rigidity of D defined as

$$S := \int_D |\nabla u|^2 \, dx.$$

There are a number of extensions of the elastic torsion problem such as the elastic-plastic torsion problem or the torsional creep problem (see Section 7.2.1). We refer the interested reader to the papers of Ting (1971), Payne and Philippin (1977a), and papers cited therein.

1.2.2 Surface of Constant Mean Curvature

In differential geometry the following problem is of interest: let H be the mean curvature of the nonparametric surface $x^{N+1} = u(x, \ldots, x^N) := u(x)$ defined over D, where the value of $u(x)$ is prescribed on ∂D. $u(x)$ is thus the solution of

$$\operatorname{div}\left(\frac{\nabla u}{\sqrt{1 + |\nabla u|^2}}\right) := \left(\frac{u_{,i}}{\sqrt{1 + u_{,i}u_{,i}}}\right)_{,i} = -NH \qquad \text{in} \quad D, \tag{1.3}$$

u given on ∂D.

This problem has been studied by a number of authors (see Section 7.2.2). Besides existence and uniqueness questions in problem (1.3), it is of interest to obtain some a priori relations among different quantities such as the surface area, the area of D, and the maximum value of u in D. Such relations in the form of inequalities will be derived in Chapters 7 and 11.

1.2.3 Capillary Surface in a Gravitational Field

D is now the cross section of a cylindrical tube. Then the equation of the surface of a liquid rising in the tube may be given in nonparametric form as $u = u(x)$, where u is the solution of

$$\operatorname{div}\left(\frac{\nabla u}{\sqrt{1+|\nabla u|^2}}\right) = \frac{(\rho-\rho_0)g}{\sigma}\,u \qquad \text{in} \quad D \subset E^2, \tag{1.4}$$

$$\frac{1}{\sqrt{1+|\nabla u|^2}}\frac{\partial u}{\partial n} = \cos\gamma \qquad \text{on} \quad \partial D, \tag{1.5}$$

where $\partial u/\partial n = \nabla u \cdot \mathbf{n}$ ($\mathbf{n}$ = outward directed unit normal vector on ∂D), and where ρ is the density of the liquid, ρ_0 the density of the gas outside, g the gravitational acceleration, σ the surface tension, and γ the wetting angle, i.e., the angle between the axis of the tube and the surface.

Again, a priori inequalities relating different quantities of interest will be derived in Chapter 7.

1.2.4 Nonlinear Eigenvalue Problems

Various models in physics, chemistry, and biology (see Section 1.3) lead to a nonlinear parabolic equation of the form

$$u_{,t} = \delta\,\Delta u + cf(u) \qquad \text{in} \quad D \times (0, T),$$

supplemented with boundary and initial conditions. Here δ and c are positive constants. A very important question is then whether or not there exists a steady state, i.e., a solution of

$$0 = \delta\,\Delta u + cf(u) \qquad \text{in} \quad D,$$

satisfying the same boundary conditions. Writing the last equation as

$$\Delta u + \lambda f(u) = 0 \qquad \text{in} \quad D \qquad \left(\lambda = \frac{c}{\delta}\right), \tag{1.6}$$

the existence problem is frequently posed as a nonlinear eigenvalue problem. By this we mean the following problem: for which positive values of λ does there exist a solution (mostly positive as well) of (1.6) satisfying given boundary conditions? In a large class of such nonlinear eigenvalue problems the following situation occurs: there is an interval (λ_*, λ^*) called the spectrum of (1.6) so that (1.6) supplemented with some boundary conditions has a solution for $\lambda \in (\lambda_*, \lambda^*)$, but no solution otherwise. Of importance are, of course, the critical values λ_* and λ^*. They depend on the given nonlinearity, the boundary conditions imposed, and the geometry of D. An exceptional case is $f(u) = u$, when $\lambda_* = \lambda^*$, i.e., the spectrum (for positive solutions) degenerates to a single value, the first eigenvalue λ_1 of

$$\Delta u + \lambda u = 0 \qquad \text{in} \quad D, \tag{1.7}$$

u satisfying some boundary conditions. Problem (1.7) is of great importance in the context of diffusion problems, and since it describes, e.g., the vibration of a membrane spanned over a region D. If the membrane is fixed on ∂D, the boundary condition becomes $u = 0$ on ∂D. The first eigenvalue λ_1 is then proportional to the square of the principal frequency of the membrane.

1.3 EXAMPLES OF PROBLEMS LEADING TO PARABOLIC EQUATIONS

1.3.1 A Single Species Population in a Finite Habitat

Let $u(x, t)$ stand for the "concentration" of some biological species, i.e., $u(x, t)$ is the number of species per unit area living at time t in the neighborhood of the point x in the finite habitat D (a domain in E^1, E^2, or E^3).

Usually one assumes that the population disperses in D in a random motion similar to the Brownian motion in physics. The species is assumed to reproduce itself at a net rate that, in general, will depend on the ambient concentration itself. Hence, in an aribtrary neighborhood $\mathcal{N}$ of a point x in D, the concentration changes according to the equation

$$\frac{\partial}{\partial t} \int_{\mathcal{N}} u \, dx = \oint_{\partial \mathcal{N}} \delta \frac{\partial u}{\partial n} \, d\sigma + \int_{\mathcal{N}} f(u) \, dx. \tag{1.8}$$

Here, the first term on the right-hand side of (1.8) is due to the diffusion flux through the boundary $\partial \mathcal{N}$ with diffusion coefficient δ, and the second term gives the net rate at which the species reproduce. Using the well-known divergence theorem and the fact that (1.8) must hold for an arbitrary neighborhood $\mathcal{N}$ of any point x in D, we are led to

$$u_{,t} = \delta \, \Delta u + f(u) \qquad \text{in} \quad D. \tag{1.9}$$

In the following, some generalized versions of Eq. (1.9) will be considered, taking into account different exterior influences that may affect the population size.

(a) *Heterogeneous reproduction rate* In a realistic habitat the food supply may vary at different places in the habitat. Also the climate may be more or less favorable in some parts of the habitat. These and other exterior influences may have the effect that the reproduction rate takes the form $f(u, x)$ instead of $f(u)$.

(b) *Heterogeneous diffusion and convection* The speed of propagation for the species may be affected by all sorts of geographical obstacles (e.g., mountains). At a certain point x in a homogeneous habitat, an observer would see a flux of species given by $\boldsymbol{\phi} = \delta \nabla u(x, t)$. But if the terrain around x is very uneven, the flux may take a different direction (and magnitude) given by $\phi^i = \delta^{ij}(x)u_{,j}$. Here δ^{ij} could be called the "diffusion tensor." It seems sensible to assume that the flux $\boldsymbol{\phi}$ cannot take a direction such that $\boldsymbol{\phi} \cdot \nabla u \leq 0$. Also, if $|\nabla u| > 0$, then $|\boldsymbol{\phi}| > 0$ should follow. Stated differently, we require that

$$\delta^{ij}(x)\xi_i\xi_j \geq \delta_0 \xi_i \xi_i \tag{1.10}$$

for some $\delta_0 > 0$ and any vector $\boldsymbol{\xi}$ at every point x of D. Hence, the operator

$$Lu := (\delta^{ij}(x)u_{,i})_{,j} \tag{1.11}$$

is uniformly elliptic in D.

In many cases the diffusion depends also on the concentration of the species. Last, there may be, e.g., a crosswind (or a stream) so that the flow of species gets a contribution in the direction of the wind. Incorporating all effects mentioned so far, we can rewrite the equation corresponding to (1.9) as

$$u_{,t} = (\delta^{ij}(u, x)u_{,j})_{,i} + v^i(x, t)u_{,i} + f(u, x). \tag{1.12}$$

In order to determine the history of the population, we also have to specify initial and boundary conditions. We assume that at a given time, say $t = 0$, the concentration is known:

$$u(x, 0) = u_0(x). \tag{1.13}$$

Let us look at three different types of boundary conditions. A first possibility is that the species cannot leave their habitat (e.g., an island). This means that the flux $\boldsymbol{\phi}$ in the normal direction (directed outward) vanishes. Thus the boundary condition becomes

$$\delta^{ij}u_{,j}n_i = 0 \qquad \text{on} \quad \partial D. \tag{1.14}$$

In the simpler case of Eq. (1.9), the boundary condition (1.14) can be written as

$$\partial u/\partial n = 0 \qquad \text{on} \quad \partial D. \tag{1.15}$$

Frequently, it is also assumed that the environment outside D is completely hostile so that no species could survive there. This is equivalent to requiring that

$$u = 0 \qquad \text{on} \quad \partial D, \tag{1.16}$$

as a second possible boundary condition (Dirichlet boundary condition).

The third possibility to be considered is that the conditions for survival outside D may gradually worsen. This could be modeled by saying that at a distance proportional to $1/h$ in the normal direction from a point on ∂D, no species can survive (see, e.g., Murray and Sperb, 1981b). Hence we require that

$$\partial u/\partial n + hu = 0 \qquad \text{on} \quad \partial D, \tag{1.17}$$

in the case described by Eq. (1.9). The parameter h, which may depend on the point on ∂D, could be called the "hostility factor." For $h \to \infty$, (1.17) leads back to (1.16).

1.3.2 Chemical Reaction Coupled with Diffusion

In order to get a chemical reaction started, the reacting species often have to be mixed first. This mixing process is in many cases due to diffusion. Consider the following reaction scheme involving two substances A and B, which react to form a product P.

$$A + nB \xrightarrow{k} P,$$

in which k is the rate for this one-sided reaction. Let u be the concentration of A, and v that of B. We may think, e.g., that the two substances are dissolved in water and simultaneously diffuse with diffusion coefficients δ_A and δ_B, respectively, and react. This process can then be modeled mathematically by the system

$$\begin{aligned} u_{,t} &= \delta_A \, \Delta u - kuv \\ v_{,t} &= \delta_B \, \Delta v - nkuv \end{aligned} \qquad \text{in} \quad D, \tag{1.18}$$

where D is the reaction vessel. Of course (1.18) will have to be complemented by initial and boundary conditions.

Analogously, a diffusion reaction process involving l substances S_i of concentrations u^i would lead to a system

$$u^i_{,t} = \delta^i \, \Delta u^i + f^i(u^1, u^2, \ldots, u^l) \qquad \text{in} \quad D, \qquad i = 1, \ldots, l. \tag{1.19}$$

Here the δ^i are the diffusion coefficients and the functions f^i describe the reaction kinetics. Equations (1.19) can also be interpreted biologically as a description of the interaction of l different species forming an ecological system. Some basic inequalities in parabolic problems of the kind described in Section 1.3 will be derived in Chapters 9 and 10.

1.4 SOME REMARKS ON A PRIORI INEQUALITIES

In the theory of partial differential equations a priori inequalities undoubtedly play a fundamental role. In many problems the derivation of certain types of a priori inequalities is a crucial step in existence and uniqueness proofs. Such a priori inequalities frequently have to hold for rather large classes of functions such as, e.g., any function of class $C^2(D)$ vanishing on the boundary ∂D.

The main point in such inequalities is that there does exist a bound of a particular type. As an illustration, consider the following example. Let u be any function of class C^1 in a plane domain D vanishing on ∂D. Then the inequality of Friedrichs (1928) states that there exists a constant $c(D)$ (independent of u!) such that the inequality

$$\int_D |\nabla u|^2\,dx \geq c(D) \int_D u^2\,dx \tag{1.20}$$

holds. The essential features of this inequality are

(a) The integral of $|\nabla u|^2$ for any $C^1(D)$ function vanishing on ∂D can be bounded in terms of the integral of u^2, and

(b) The constant c appearing in (1.20) depends only on the domain.

It would be easy to show by counterexamples that there is no constant $c(D)$ such that the inequality sign in (1.20) could be reversed.

For many theoretical considerations it will be unimportant to know more about the constant $c(D)$ than its existence. However, a natural question arising in this context is certainly the following: what is the best possible constant $c(D)$ in (1.20); or equivalently, what is

$$\inf_{\substack{v \in C^1(D) \\ v=0 \text{ on } \partial D}} \int_D |\nabla v|^2\,dx \Big/ \int_D v^2\,dx? \tag{1.21}$$

Of course it is well known that the optimal value $c(D)$ is given by the first eigenvalue λ_1 of

$$\Delta\phi + \lambda\phi = 0 \quad \text{in} \quad D, \qquad \phi = 0 \quad \text{on} \quad \partial D. \tag{1.22}$$

Yet this last information may not be explicit enough for many practical applications. The inequality of Faber and Krahn (see Chapter 11) states that for a simply connected plane domain one has

$$\begin{aligned} &\lambda_1 \geq j_0^2\pi/A. \\ &j_0^2 = \text{first zero of the Bessel function } j_0, \\ &j_0 \cong 2.405, \\ &A = \text{area of } D. \end{aligned} \tag{1.23}$$

The important points in inequality (1.23) are now

(a) λ_1 can be bounded in terms of the area of D; and

(b) inequality (1.23) is optimal in the sense that there exists a domain (the circle in this case) for which the equality sign holds in (1.23). Such an inequality is called "isoperimetric" in analogy or the "classical isoperimetric inquality," $L^2 \geq 4\pi A$, between the area A of a domain and the length L of its perimenter.

Using (1.23), the a priori inequality (1.20) could be written in more explicit form as

$$\int_D |\nabla u|^2 \, dx \geq \frac{18.16}{A} \int_D u^2 \, dx. \tag{1.20'}$$

An essential point in an isoperimetric inequality is the knowledge of the geometrical quantities are involved. So in our example (1.23), we have seen that if the area of a simply connected plane domain is known, a lower bound for the eigenvalue λ_1 can be given. If instead of problem (1.22), we consider the "free membrane eigenvalue problem"

$$\Delta v + \mu v = 0 \quad \text{in} \quad D, \qquad \frac{\partial v}{\partial n} = 0 \quad \text{on} \quad \partial D, \tag{1.24}$$

then it can be shown by counterexamples (see, e.g., Section 6.2.3) that for the first nonzero eigenvalue μ_1 in problem (1.24) no bound of the form

$$\mu_1 \geq \text{number}/A \tag{1.25}$$

can exist. It is therefore of great interest to know if, e.g., an eigenvalue can be bounded by a certain geometrical quantity. Let us mention another example: let ρ be the radius of the largest circle inscribed in a plane domain D. It is then easy to see that in general it is not possible to have a bound of the type

$$\lambda_1 \geq \text{number}/\rho^2$$

in problem (1.22). However, if we restrict our attention to *convex* domains, we have (see, e.g., Chapter 8)

$$\lambda_1 \geq \pi^2/4\rho^2 \cong 2.47/\rho^2, \tag{1.26}$$

and this last inequality is again isoperimetric in the sense that the equality sign holds if D is an infinite strip of width 2ρ. Thus, isoperimetric inequalities may give important information about how geometrical shape affects the magnitude of certain quantities of interest in partial differential equations.

There is also a practical advantage in having isoperimetric inequalities. In most problems involving partial differential equations explicit solutions and numbers for quantities of interest are known in only a few cases. Often it is possible to obtain bounds for a certain quantity based on some general comparison principle. Then isoperimetric inequalities may lead to improved bounds. As an illustration, consider the torsion problem for a square S of side a:

$$\Delta u + 1 = 0 \quad \text{in} \quad S, \qquad u = 0 \quad \text{on} \quad \partial S. \tag{1.27}$$

In a number of applications (see e.g., Chapter 3) it is important to know an upper bound for $u_M := \max_S u$. From the well-known monotonicity of u with respect to the domain, it follows that the value of u_M for a square is bounded above by the corresponding value for the circumscribed circle. Hence we have

$$u_M \leq a^2/8. \tag{1.28}$$

On the other hand, an isoperimetric inequality of Pólya and Szegö (see Chapter 11) states that the value of u_M is bounded above by the corresponding value of a circle of the same area as S, i.e.,

$$u_M \leq a^2/4\pi, \tag{1.29}$$

which is a considerable improvement of (1.28).

A large number of similar examples could easily be given. Other aspects regarding this subject are discussed in the review article of Payne (1967) and in the books of Pólya and Szegö (1951) as well as in Bandle (1980).

In this book we focus our attention on obtaining isoperimetric bounds in various problems. The main idea consists of first deriving an optimal maximum principle for some functional of the solution from which isoperimetric inequalities will follow.

Chapter 2

MAXIMUM PRINCIPLES IN ELLIPTIC AND PARABOLIC PROBLEMS

The main tools used in this book are the maximum principles in elliptic and parabolic problems. A thorough treatment of this important topic would easily fill a book. Therefore we shall restrict our attention to the maximum principles that will actually be used later. Proofs are presented only in a few simple cases, which nevertheless show the main ideas. For readers interested in more details as well as further applications and extensions, see the excellent book of Protter and Weinberger (1967).

2.1 INTRODUCTION: ONE-DIMENSIONAL CASE

Let $u(x)$ be a twice continuously differentiable function of the single variable x, for x in the interval $I = (a, b)$. Suppose $u(x)$ satisfies the differential inequality

$$u''(x) \geq 0. \tag{2.1}$$

In Figure 2.1 a sketch of possible graphs is given. Graph (1) represents the case that $u''(x) \geq 0$ and $u''(x) \not\equiv 0$, whereas graphs (2) and (3) belong to the special case $u'' \equiv 0$. The important properties of u that are shown in these graphs are

(i) $u(x)$ cannot have an interior maximum in I, unless $u(x) \equiv \text{const}$.

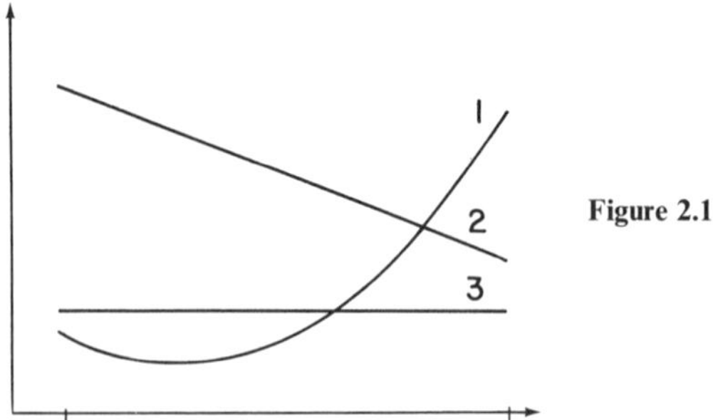

Figure 2.1

(ii) At a point on ∂I (i.e., $x = a$ or $x = b$) where $u(x)$ attains its maximum, the slope of u is nonzero. More precisely, $du/dn > 0$ at such a point. Here, d/dn denotes the outward derivative on ∂I, i.e.,

$$\left.\frac{du}{dn}\right|_{x=a} := -u'(a), \qquad \left.\frac{du}{dn}\right|_{x=b} := u'(b).$$

The only exception is again the trivial case $u(x) \equiv \text{const}$ when $du/dn = 0$.

Of course, drawing a graph is by no means a valid proof. In the next step, let us consider the functions $u(x)$ satisfying the strict inequality

$$u''(x) + b(x)u'(x) > 0 \qquad \text{in} \quad I. \tag{2.2}$$

Here, $b(x)$ is a bounded function in I. We can easily prove that $u(x)$ cannot have an interior maximum in I; at such a point, we would find $u'' > 0$, because of (2.2). This is obviously a contradiction since at a maximum we must have $u''(x) \leq 0$. The main difficulty in the proof of the maximum principles arises when the strict inequality such as in (2.2) is replaced by a nonstrict one. The usual proofs then reduce the nonstrict inequality to a strict one by introducing an auxilliary function. This procedure is demonstrated in the proof of

Theorem 2.1 (One-dimensional weak maximum principle) Let u be a nonconstant function satisfying

$$Lu := u'' + b(x)u' \geq 0 \qquad \text{in} \quad I, \tag{2.3}$$

with b bounded in I. Then, a nonconstant u can assume its maximum only on ∂I.

Proof We assume that u takes its maximum M at $x = c$, $a < c < b$, and that there is a point $x = d$, $c < d < b$, where $u < M$. We now introduce

$$z(x) = e^{\alpha(x-c)} - 1$$

with a positive constant α that is to be determined later. A simple computation gives

$$Lz = \alpha(\alpha + b(x))e^{\alpha(x-c)}.$$

Since $b(x)$ is bounded we can choose α such that

$$Lz > 0 \quad \text{in} \quad (a, d).$$

Next we set

$$w(x) = u(x) + \varepsilon z(x).$$

Here the constant ε is chosen such that

$$\varepsilon < \frac{M - u(d)}{z(d)}.$$

It is now easy to check that

$$\begin{aligned} w(x) &< M \quad \text{in} \quad (a, c), \\ w(c) &= M, \\ w(d) &< M. \end{aligned}$$

This means that w has a maximum $\bar{M} \geq M$ in the interior of (a, d). But

$$Lw = Lu + \varepsilon Lz > 0 \quad \text{in} \quad (a, d),$$

which is impossible as we have seen in the discussion following the strict inequality (2.2). Hence, u must attain its maximum on ∂I.

The graphs in Fig. 2.1 showed the second important property of a function u satisfying the differential inequality (2.1): $du/dn > 0$ at the boundary point where the maximum is attained. The next result is an extension to the case in which u satisfies the inequality (2.3) instead of (2.1).

Theorem 2.2 (One-dimensional strong maximum principle) Suppose the nonconstant function u satisfies

$$Lu := u'' + b(x)u' \geq 0 \quad \text{in} \quad I,$$

with b bounded. Then, u takes its maximum on ∂I and $du/dn > 0$ at such a point.

Proof Again, the essential part is to show the strict inequality since $du/dn \geq 0$ is certainly true. We suppose that $u(a) = M$ and $u(x) \leq M$ in I. Also, $u(d) < M$ for some point $d \in (a, b)$. As before, we set

$$z(x) = e^{\alpha(x-a)} - 1,$$

and choose α such that $Lz > 0$ in (a, d). Analogously we take

$$w(x) = u(x) + \varepsilon z(x),$$

choosing ε so that

$$\varepsilon < \frac{M - u(d)}{z(d)}.$$

Then we have $Lw > 0$ in (a, d) and the maximum of w must therefore occur at $x = a$ or at $x = d$. But $w(a) = M > w(d)$ because of our choice of ε. Therefore, we certainly have

$$w'(a) = u'(a) + \varepsilon z'(a) = u'(a) + \varepsilon\alpha \leq 0,$$

which in turn implies that

$$\left.\frac{du}{dn}\right|_{x=a} := -u'(a) > 0.$$

The procedure is similar if $u(b) = M$.

Remarks on Theorems 2.1 and 2.2 (a) If we apply Theorems 2.1 and 2.2 to the function $-u$, we find that a nonconstant function u satisfying

$$Lu \leq 0 \qquad \text{in} \quad I$$

must assume its minimum on ∂I. Furthermore, at the point of ∂I where the minimum is attained, we have $du/dn < 0$. In particular, we see that if $Lu = 0$ in I, then the maximum and minimum of u must occur on ∂I. In view of the extensions that will be discussed later on, we also note that if

$$Lu = 0 \qquad \text{in} \quad I \qquad \text{and} \qquad u(a) = u(b)$$

or

$$u'(a) = u'(b) = 0,$$

then $u \equiv \text{const}$.

(b) It follows from the proof of Theorem 2.1 that $b(x)$ need only be bounded in closed subintervals of I, and bounded from below if $u(a) = M$, or bounded from above if $u(b) = M$. However, if $b(x)$ is not bounded in I, Theorem 2.1 or 2.2 does not hold in general. A counterexample (see Protter and Weinberger, 1967) is given by a solution of

$$u'' + b(x)u' = 0$$

with

$$b(x) = \begin{cases} 3/x & \text{for} \quad x \neq 0, \\ 0 & \text{for} \quad x = 0. \end{cases}$$

A solution is, e.g., $1 - x^4$, showing that Theorem 2.1 would not hold on any interval containing $x = 0$ in its interior. On the other hand, Theorem 2.2 is violated if $x = 0$ is an end point, since $u'(0) = 0$.

(c) It also follows from the proofs that Theorems 2.1 and 2.2 still hold for functions u satisfying

$$u'' + b(x, u)u' \geq 0 \qquad \text{in} \quad I,$$

provided that $b(x, \xi)$ is a bounded function of its two arguments for $x \in I$ and $\xi \in (-\infty, \infty)$.

In applications, a somewhat extended form of Theorems 2.1 and 2.2 is often needed. Suppose u satisfies

$$Lu + hu = u'' + b(x)u' + h(x)u \geq 0 \qquad \text{in} \quad I. \tag{2.4}$$

Then the question arises as to whether or not it is still possible to prove similar maximum principles. The simple example $u(x) = \sin x$, i.e., when u satisfies ($h \equiv 1$)

$$u'' + u = 0$$

shows that $h(x)$ must be restricted.

Analogously, as before, one can now prove

Theorem 2.3 Let u be a nonconstant solution of $u'' + b(x)u' + h(x)u > 0$ in I, where b and h are bounded functions in I, and $h \leq 0$. Then a nonnegative maximum of u can only occur on ∂I, and $du/dn > 0$ there.

Proof The essential part of the proof is again the fact that the constant α of $z(x) := e^{\alpha(x-c)} - 1$ can be chosen such that

$$e^{-\alpha(x-c)}(L + h)z = \alpha^2 + \alpha b + h(1 - e^{-\alpha(x-c)}) \geq 0.$$

More details can again be found in Protter and Weinberger (1967).

Remarks (a) The function u can assume a *negative* maximum in an interval I as the following example shows: Take

$$u(x) = -1 - x^2 \qquad \text{and} \qquad I = (-\tfrac{1}{2}, +\tfrac{1}{2}).$$

Then we have

$$u'' + xu' - 2u = 0.$$

Clearly, u assumes a negative maximum in the interior of I.

(b) Applying Theorem 2.3 to $-u$, we conclude that if

$$u'' + b(x)u' + h(x)u \leq 0$$

with the same restrictions on b, h as before, then u must assume a non-positive minimum on ∂I and $du/dn < 0$ there.

(c) If u satisfies

$$u'' + b(x)u' + h(x)u = 0,$$

b, h as before, and $u = 0$ on ∂I, it thus follows that $u \equiv 0$ in I.

2.2 ELLIPTIC PROBLEMS

Let D be a finite domain in n-dimensional Euclidean space E^n. The results of Section 2.1 can be extended to the case in which the operator L takes the form

$$Lu := a^{ij}(x)u_{,ij} + b^i(x)u_{,i}, \tag{2.5}$$

where $a^{ij} = a^{ji}$ and L is assumed to be uniformly elliptic, i.e.,

$$a^{ij}(x)\xi_i\xi_j \geq \mu_0\xi_i\xi_i \qquad \text{in} \quad D \tag{2.6}$$

for any vector $\boldsymbol{\xi}$ and some positive constant μ_0. Again, it is not hard to check that if a function u satisfies

$$Lu > 0 \qquad \text{in} \quad D, \tag{2.7}$$

then u cannot attain its maximum M at an interior point $\bar{x}$ of D, unless $u \equiv$ const. To see this we choose $\bar{x}$ as the origin of our coordinate system and perform an orthogonal transformation T of our coordinates:

$$T\colon \quad y^k = t_i^k x^i.$$

Then, since

$$\frac{\partial^2 u}{\partial x^i \partial x^j} = t_i^k t_j^l \frac{\partial^2 u}{\partial y^k \partial y^l},$$

we may select the transformation T such that

$$a^{ij}(\bar{x}) \frac{\partial^2 u}{\partial x^i \partial x^j} = a^{ij}(\bar{x}) t_i^k t_j^l \frac{\partial^2 u}{\partial y^k \partial y^l} = \mu_k(\bar{x}) \frac{\partial^2 u}{(\partial y^k)^2},$$

i.e., the transformed matrix is diagonal. It is well known from linear algebra that the $\mu_k(\bar{x})$ are the eigenvalues of $a^{ij}(\bar{x})$, and by (2.6), $\mu_k(\bar{x}) \geq \mu_0 > 0$. On the other hand, we know that at the maximum we must have

$$u_{,ii} \leq 0 \qquad \text{for all} \quad i = 1, \ldots, n \qquad \text{(no sum).}$$

Hence, at $\bar{x}$ we have

$$Lu = \mu_k(\bar{x})u_{,kk} \leq 0,$$

in contradiction to (2.7). If u satisfies only the nonstrict inequality

$$Lu \geq 0, \tag{2.8}$$

one can again choose a similar auxiliary function $z(x)$ as in Section 2.1. Reasoning analogous to that for the one-dimensional case then leads to the extension of Theorem 2.1, which may be stated as

Theorem 2.4 Suppose u satisfies the inequality

$$Lu = a^{ij}(x)u_{,ij} + b^{i}(x)u_{,i} \geq 0$$

in some finite domain $D \subset E^n$, and the coefficients of L are bounded in D. Then, u cannot assume its maximum at an interior point of D unless $u \equiv$ const.

For the proof we refer once again to Protter and Weinberger (1967).

The strong maximum principle also has an extension to the higher-dimensional case. A new feature in the proof is that the boundary ∂D must have the "interior sphere property" at a point x_M of ∂D where the maximum is attained (see Figure 2.2). There must be a sphere S of some radius $r_0 > 0$ contained in D such that $S \cap \partial D = \{x_M\}$.

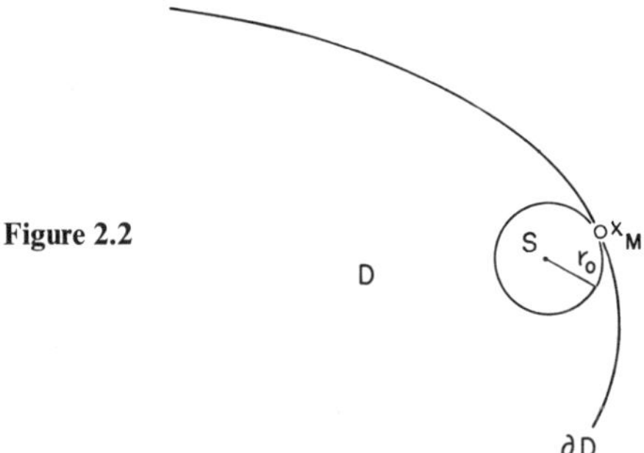

Figure 2.2

Again, the ideas used in the proof of the one-dimensional strong maximum principle can be adapted for the higher-dimensional case and lead to

Theorem 2.5 (Strong maximum principle of E. Hopf) Let u be a nonconstant solution of $Lu \geq 0$ in the bounded finite domain $D \subset E^n$. Suppose u attains its maximum M at some point x_M of ∂D, and ∂D has the interior sphere property at x_M. Let $\mathbf{v}$ be any vector pointing outward at x_M. Then, $\partial u/\partial v > 0$.

The details of the proof can be found in Protter and Weinberger (1967, pp. 65–66). Clearly, one can add yet another extension by considering functions u satisfying the inequality

$$Lu + h(x)u \geq 0 \quad \text{in} \quad D, \tag{2.9}$$

with L as in (2.8) and a bounded function $h(x)$ satisfying

$$h(x) \leq 0 \quad \text{in} \quad D. \tag{2.10}$$

As a final result in this section, we state the following theorem.

Theorem 2.6 The conclusions of Theorems 2.4 and 2.5 also hold for solutions of inequality (2.9) provided (2.10) holds, and $M \geq 0$.

Remarks (a) The strong maximum principle will in general not be valid if the interior sphere property is not satisfied. As an example, take for D the square

$$Q := \{x, y \mid -\tfrac{1}{2}\pi < x < \tfrac{1}{2}\pi,\ -\tfrac{1}{2}\pi < y < \tfrac{1}{2}\pi\}$$

and

$$u(x, y) = -\cos x \cos y.$$

Then,

$$\Delta u = 2 \cos x \cos y \geq 0 \quad \text{in} \quad Q,$$

and u assumes its maximum $M = 0$, e.g., at the corner $(\tfrac{1}{2}\pi, \tfrac{1}{2}\pi)$. But there $u_{,x} = 0$ and $u_{,y} = 0$, so that any outwardly directed derivative $\partial u/\partial v$ also vanishes. Note, however, that for any other point on the segment

$$\{x = \tfrac{1}{2}\pi,\ -\tfrac{1}{2}\pi < y < \tfrac{1}{2}\pi\},$$

where the maximum $M = 0$ is also attained, we have

$$u_{,x} > 0, u_y = 0,$$

implying that $\partial u/\partial v > 0$ there.

(b) Replacing u by $-u$ in Theorems 2.4–2.6, we find the corresponding minimum principles as in the previous section. Note that $h(x) \leq 0$ is still required, but $M \leq 0$ if $h(x) \not\equiv 0$.

(c) The maximum principles also hold for solutions of

$$a^{ij}(x, u, u_{,k})u_{,ij} + b^i(x, u, u_{,k})u_{,i} + h(x, u, u_{,k})u \geq 0 \tag{2.11}$$

provided that a^{ij}, b^i, and h are bounded functions of their arguments; $h \leq 0$ and $a^{ij}(x, \xi, \eta)$ is uniformly elliptic as before for $-\infty < \xi, \eta < \infty$ and $x \in D$.

There are a number of other extensions of the maximum principles for which I refer the reader to the book of Protter and Weinberger (1967). In this book, I shall only apply the maximum principles as stated in Theorems 2.1–2.6 and the results of the following section.

(d) The condition that $h(x) \le 0$ can often be relaxed. Let ϕ be the first (positive) eigenfunction of

$$\Delta\phi + \lambda_1(\varepsilon)\phi = 0 \qquad \text{in} \quad D,$$

$$\varepsilon\frac{\partial\phi}{\partial n} + \phi = 0 \qquad \text{on} \quad \partial D,$$

and let $\varepsilon > 0$. It is well known that if $0 < \varepsilon < \infty$, one has $\phi > 0$ in $\bar{D}$. Suppose now that the function u satisfies

$$\Delta u + h(x)u \ge 0 \qquad \text{in} \quad D, \qquad u \le 0 \qquad \text{on} \quad \partial D,$$

where the function $h(x)$ only satisfies

$$h(x) \le \lambda_1(\varepsilon)$$

for some $\varepsilon > 0$. Theorem 2.6 does not allow us to conclude that $u \le 0$ must hold in D. Nevertheless we can prove that $u \le 0$ in D as follows: Set $v = \phi v$, for some function $v(x)$. Then we have

$$\phi\,\Delta v + 2\,\nabla\phi\cdot\nabla v + v\,\Delta\phi + h\phi v = \phi\,\Delta v + 2\,\nabla\phi\cdot\nabla v + (h - \lambda_1(\varepsilon))\,\phi v \ge 0,$$

and since $\phi > 0$ for $\varepsilon > 0$, it follows that v satisfies

$$\Delta v + b^i v_{,i} + \hat{h}v \ge 0 \qquad \text{in} \quad D, \qquad v \le 0 \qquad \text{on} \quad \partial D.$$

Now we can apply Theorem 2.1, which tells that $v \le 0$ in D, and therefore $u \le 0$, if $h(x) \le \lambda_1(\varepsilon)$ and $\varepsilon > 0$. Letting $\varepsilon \to 0$ it follows by continuity that still $u \le 0$ if $h \le \lambda_1$, where $\lambda_1 = \lambda_1(0)$.

(e) If the condition $h(x) \le \lambda_1$ is violated, then the following antimaximum principle holds (see Clement and Peletier, 1970): Suppose that

$$Lu + \lambda u = a^{ij}(x)u_{,ij} + b^i u_{,i} + \lambda u < 0 \qquad \text{in} \quad D,$$

$$(\partial u/\partial N) + \sigma u = 0 \qquad \text{on} \quad \partial D,$$

and denote by λ_1 the first eigenvalue of

$$L\phi + \lambda_1\phi = 0 \qquad \text{in} \quad D, \qquad (\partial u/\partial N) + \sigma\phi = 0 \qquad \text{on} \quad \partial D.$$

Here, $\sigma > 0$ and $\partial/\partial N$ denotes the conormal derivative, i.e.,

$$\partial\psi/\partial N = a^{ij}u_{,i}n_j, \qquad \mathbf{n} = \text{outward normal on} \quad \partial D.$$

Then, there exists a $\delta > 0$, such that if

$$\lambda_1 < \lambda < \lambda_1 + \delta,$$

one has

(i) $u < 0$ in D and

(ii) $\partial u/\partial N > 0$ at points where $u = 0$ on ∂D.

Note that by remark (d), if $\lambda \leq \lambda_1$, one would have,

(i) $u > 0$ in D,
(ii) $\partial u/\partial N < 0$ at points where $u = 0$ on ∂D.

2.3 PARABOLIC PROBLEMS

We now consider functions $u(x, t)$, where x is a point of some finite domain $D \subset E^n$ and t lies in some interval $(0, T)$. In view of our later applications we shall refer to t as "time." Assume now that the function u satisfies

$$\Delta u - u_{,t} > 0 \quad \text{in} \quad D \times (0, T). \tag{2.12}$$

Suppose $u(x, t)$ attains its maximum M for some $x_M \in D$, $t_M \in (0, T)$. At (x_M, t_M) we must have $\Delta u \leq 0$, and hence $u_{,t} < 0$ at (x_M, t_M). But if x_M is an interior point of D, this means that $t_M = 0$. Hence we see that any solution u of (2.12) must take its maximum either for $t = 0$ or on the boundary ∂D. We can also state this result more geometrically by saying that a nonconstant solution of (2.12) in the cylindrical region $R := D \times (0, T)$ (see Figure 2.3) can assume its maximum only at the "bottom" $D \times \{0\}$ or on the "sidewalls" $\partial D \times [0, T]$.

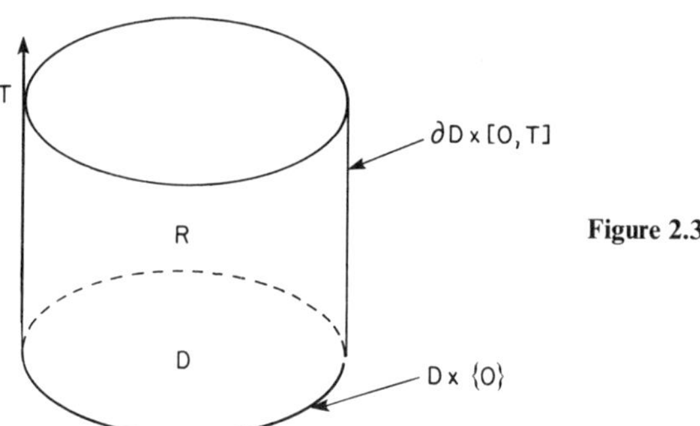

Figure 2.3

In analogy to Section 2.2, it is quite obvious that the Laplacian in inequality (2.12) can be replaced by a uniformly elliptic operator L. Furthermore, one needs only a nonstrict inequality. For the proofs of the following three theorems we also refer to Protter and Weinberger (1967: pp. 174–175).

Theorem 2.7 Let u be a nonconstant solution of

$$Lu - u_{,t} = a^{ij}(x, t)u_{,ij} + b^i(x, t)u_{,i} - u_{,t} \geq 0 \quad \text{in} \quad D \times (0, T), \tag{2.13}$$

D a finite domain in E^n, $T < \infty$, and L a uniformly elliptic operator with bounded coefficients a^{ij} and b^i. Then u can attain its maximum only for $t = 0$ or on ∂D.

The strong maximum principle for elliptic problems has also been extended to parabolic problems by Friedman (1964). For our purposes, we state it as

Theorem 2.8 Assume that the conditions of Theorem 2.7 are satisfied. If u attains its maximum M at some point x_M of $\partial D \times (0, T)$, where ∂D has the interior sphere property, then $\partial u/\partial v > 0$ at x_M.

Finally, both of our preceding results have extensions to the case in which u satisfies

$$Lu + h(x,t)u - u_{,t} = a^{ij}(x,t)u_{,ij} + b^i(x,t)u_{,i} + h(x,t)u - u_{,t} \geq 0 \quad (2.14)$$

in $D \times (0, T)$. Again, we suppose that $h(x, t) \leq 0$ in $D \times (0, T)$.

Theorem 2.9 The conclusions of Theorems 2.7 and 2.8 remain valid for nonconstant solutions of (2.14), provided that $M \geq 0$.

Remarks (a) If u satisfies (2.14), $h(x, t)$ is bounded from above and $M = 0$, we can set $v(x, t) = ue^{-\alpha t}$. Then v satisfies

$$e^{\alpha t}\{Lv + (h - \alpha)v - v_{,t}\} \geq 0. \quad (2.15)$$

If h is bounded from above, we can choose α large enough so that $h - \alpha \leq 0$ in $D \times (0, T)$, and hence Theorem 2.9 is applicable. In particular, it follows that $u(x, t) \leq 0$ in $D \times (0, T)$ if $u(x, 0) \leq 0$ and $u(x, t) \leq 0$ for $x \in \partial D$. It will be an important advantage in parabolic problems that the restriction $h \leq 0$ is not necessary whenever $M = 0$.

(b) By changing from u to $-u$ in Theorems 2.7–2.9, we obtain the corresponding minimum principles.

(c) Again the coefficients of L and h may also depend on the function u and its derivatives $u_{,k}$ provided the ellipticity of L, the boundedness of a^{ij}, b^i, h and $h \leq 0$ are still guaranteed.

2.4 PARABOLIC SYSTEMS

Let us first consider a set of functions u^μ, $\mu = 1, 2, \ldots, m$, satisfying a weakly coupled system of parabolic inequalities of the form

$$\Delta u^\mu + \sum_{\nu=1}^{m} h_{\mu\nu} u^\nu - u^\mu_{,t} > 0 \qquad \text{in} \quad D \times (0, T), \qquad \mu = 1, \ldots, m, \quad (2.16)$$

with $h_{\mu\nu} = h_{\mu\nu}(x, t)$ as the coupling term. Of course, we cannot expect that all the functions u^μ assume their maximum at the same "point" (x_M, t_M). We therefore look for a somewhat weaker form of a maximum principle. Suppose that for $\mu = 1, \ldots, m$, we have

$$u^\mu(x, 0) < 0, \qquad u^\mu(x, t) < 0 \qquad \text{if} \quad x \in \partial D. \tag{2.17}$$

Does $u^\mu(x, t) < 0$ in $D \times (0, T)$ follow? To find out, suppose contrariwise that for some fixed $\mu = \mu_0$ we have $u^{\mu_0}(x_0, t_0) \geq 0$ for some $(x_0, t_0) \in D \times (0, T)$. Since u^{μ_0} is initially nonpositive, assume that $\bar{t} \leq t_0$ is the first time when $u^{\mu_0} = 0$. Hence,

$$u^{\mu_0}_{,t}(x_0, \bar{t}) \geq 0 \tag{2.18}$$

must hold. Clearly, u^{μ_0} has a maximum $M = 0$ in $D \times [0, \bar{t}]$ at x_0 in D, therefore $\Delta u^{\mu_0}(x_0, t) \leq 0$ must hold. But the μ_0th inequality gives

$$\Delta u^{\mu_0} > -\sum_{\nu=1}^{m} h_{\mu_0 \nu} u^\nu \tag{2.19}$$

at $(x_0, \bar{t})$, because of (2.18). Since $u^{\mu_0}(x_0, \bar{t}) = 0$ and $u^\nu(x_0, \bar{t}) < 0$, we see that (2.19) would lead to a contradiction if we require that

$$h_{\mu\nu}(x, t) \geq 0 \qquad \text{for} \quad \mu \neq \nu. \tag{2.20}$$

The generalization then reads as follows:

Theorem 2.10 Let the functions u^μ, $\mu = 1, \ldots, m$, satisfy the weakly coupled parabolic system

$$L^\mu u^\mu + \sum_{\nu=1}^{m} h_{\mu\nu} u^\nu - u^\mu_{,t} \geq 0 \qquad \text{in} \quad D \times (0, T), \tag{2.21}$$

with

$$L^\mu u^\mu := a^{ij}_{(\mu)}(x, t) u^\mu_{,ij} + b^i_{(\mu)}(x, t) u^\mu_{,i}, \tag{2.22}$$

uniformly elliptic operators with bounded coefficients. Suppose that $h_{\nu\mu} \geq 0$ for $\mu \neq \nu$. Then, if for $\mu = 1, \ldots, m$, $u^\mu \leq 0$ in $D \times \{0\}$ and on $\partial D \times (0, T)$, it follows that $u^\mu \leq 0$ in $D \times (0, T)$ for all μ. Furthermore, if for some $\mu = \mu_0$, we have $u^{\mu_0}(x_0, t_0) = 0$ for $(x_0, t_0) \in D \times (0, T)$, then $u^{\mu_0}(x, t) \equiv 0$ in $D \times (0, t_0)$.

A proof of Theorem 2.10 is given by Protter and Weinberger (1967: pp. 189–190). It consists mainly of reducing the nonstrict inequalities to strict ones. Again, there is a corresponding strong maximum principle.

Theorem 2.11 Under the assumptions of Theorem 2.10, suppose that $u^{\mu_0}(x_0, t_0) = 0$ for some $(x_0, t_0) \in \partial D \times (0, T)$ and that D has the interior sphere property at x_0. Then,

$$\frac{\partial u^{\mu_0}}{\partial \nu} > 0 \qquad \text{at} \quad x_0$$

for any outward directional derivative, unless $u^{\mu_0} \equiv 0$.

Remarks (a) Suppose that $u^\mu \leq M$ at $t = 0$ and on $\partial D \times (0, T)$ for all $\mu = 1, \ldots, m$ and $M > 0$. For a set of functions satisfying the system (2.21), one would like to conclude that $u^\mu \leq M$ in $D \times (0, T)$ also. However, the analogous reasoning as before applied to the functions $u^\mu := u^\mu - M$, shows that Theorems 2.10 and 2.11 remain valid with the maximum 0 replaced by $M \geq 0$ provided the additional hypothesis

$$\sum_{\nu=1}^{m} h_{\mu\nu} \leq 0 \qquad \text{for} \quad \mu = 1, \ldots, m \tag{2.23}$$

is satisfied.

(b) If the functions u^μ satisfy the system (2.21), then the functions $\hat{u}^\mu := u^\mu e^{-\alpha t}$ satisfy

$$L^\mu \hat{u}^\mu + \sum_{\nu=1}^{m} (h_{\mu\nu} - \alpha\delta_{\mu\nu})\hat{u}^\nu - \hat{u}^\mu_{,t} \geq 0 \qquad (\delta_{\mu\nu}\text{: Kronecker symbol}). \tag{2.24}$$

We can always choose α large enough (the $h_{\mu\nu}$ are bounded) so that the inequality (2.23) holds. Hence if $u^\mu \leq M$ at $t = 0$ and on $\partial D \times (0, T)$, we can conclude that

$$u^\mu \leq M e^{\alpha t} \qquad \text{in} \quad D \times (0, T), \tag{2.25}$$

where

$$\alpha = \max_{\mu} \sum_{\nu=1}^{m} h_{\mu\nu} \geq 0.$$

(c) If the functions v^μ satisfy the system of *elliptic* inequalities

$$L^\mu v^\mu + \sum_{\nu=1}^{m} h_{\mu\nu} v^\nu \geq 0 \qquad \text{in} \quad D, \tag{2.26}$$

with $h_{\mu\nu}$ satisfying (2.20) and (2.23), we can apply Theorem 2.10 and Theorem 2.11 to conclude that if

$$v^\mu \leq M \qquad \text{on} \quad \partial D, \qquad M \geq 0,$$

then

$$v^\mu \leq M \qquad \text{in} \quad D$$

follows.

(d) Obviously, one can derive the corresponding minimum principles again as before.

(e) The coefficients $a^{ij}_{(\mu)}$, $b^i_{(\mu)}$, and $h_{\mu\nu}$ may depend on the u^ν and the derivatives $u^\nu_{,k}$ as long as the corresponding hypotheses on the ellipticity of the L^μ and the $h_{\mu\nu}$ are valid.

Chapter 3

EXISTENCE AND REGULARITY RESULTS

3.1 EXISTENCE RESULTS BY THE METHOD OF SUB- AND SUPERSOLUTIONS

3.1.1 Introduction

In recent years a great number of papers have been published dealing with the existence problem for solutions to semilinear elliptic equations of the form

$$Lu + f(x,u) = 0 \qquad \text{in} \quad D, \tag{3.1}$$

$$Bu = g(x) \qquad \text{on} \quad \partial D, \tag{3.2}$$

where L is a uniformly elliptic operator as in Chapter 2, and the boundary operator is given by

$$Bu := \frac{\partial u}{\partial v} + hu, \tag{3.3}$$

with v an outward-pointing, nowhere tangent vector field on ∂D, and $h(x) \geq 0$ on ∂D. In particular, the reader is referred to the books of Miranda (1970) and Ladishenskaya and Ural'ceva (1968), and the review article of Amann (1976b), which also contains an extensive bibliography on the subject.

Numerous methods have been derived in order to prove existence in problem (3.1), (3.2) as well as various extensions of it. Often, a fixed-point algorithm in an abstract function space is used to prove existence of a solution in problem (3.1), (3.2). It would go beyond the scope of this book to discuss this topic. The reader is again referred to the literature cited previously.

Here, we shall concentrate on one particular technique that is intimately connected with the maximum principles of the previous section: the so-called method of sub- and supersolutions. The first publication in which this technique in a special case of problem (3.1), (3.2), is employed seems to be the paper by Hudjaev (1964). This method has been rediscovered and extended in various directions by different authors.

3.1.2 Existence Results in Nonlinear Elliptic Problems

Before presenting the method of sub- supersolutions let us consider the following situation. Suppose there exists a function $\underline{u}$, called a subsolution to Eqs. (3.1) and (3.2) satisfying

$$L\underline{u} + f(x, \underline{u}) \geq 0 \qquad \text{in} \quad D, \tag{3.4}$$

$$B\underline{u} \leq g(x) \qquad \text{on} \quad \partial D. \tag{3.5}$$

Assume further that for $\xi, \eta \in \mathbb{R}$ we have

$$f(x, \xi) - f(x, \eta) \geq -\gamma|\xi - \eta|$$

for all $x \in D$ and some finite $\gamma \geq 0$. Let u be a solution of Eqs. (3.1), (3.2), and set

$$w := \underline{u} - u.$$

Then w satisfies

$$Lw = L\underline{u} - Lu \geq f(x, u) - f(x, \underline{u}) \geq \gamma w,$$

i.e.,

$$Lw - \gamma w \geq 0 \qquad \text{in} \quad D,$$

and clearly

$$Bw = \partial w/\partial v + hw \leq 0 \qquad \text{on} \quad \partial D.$$

If we further assume that D has the interior sphere property, we can apply the strong maximum principle. A nonconstant function w (i.e., $\underline{u}$ is *not* a solution itself) must assume a nonnegative maximum on ∂D. But on ∂D we have

$$\partial w/\partial v \leq -hw \leq 0,$$

contradicting the strong maximum principle if $\max_{\partial D} w \geq 0$. Therefore

$$u \geq \underline{u} \quad \text{in} \quad D \tag{3.6}$$

must hold.

A supersolution $\bar{u}$ is defined by reversing the inequality signs in (3.4), (3.5). Exactly as before, one can show that

$$u \leq \bar{u} \tag{3.7}$$

would follow.

The inequalities (3.6), (3.7) tell us that, provided that a solution u of (3.1), (3.2) exists, it must lie between a sub- and a supersolution. However we do not know whether there exists a solution at all. An important result now is that the converse is indeed true, i.e., if a sub- and a supersolution exist, then there also exists a solution.

Before stating the exact result, let us look at a more general problem, namely, a solution of (3.1) satisfying

$$Bu = g(x, u) \quad \text{on} \quad \partial D. \tag{3.8}$$

Let I be a nonempty subinterval of $\mathbb{R}$ and $k(s)$ an arbitrary function of the variable $s \in \mathbb{R}$. We repeat that k is called α-Hölder-continuous if

$$|k(s_1) - k(s_2)| \leq \gamma(I)|s_1 - s_2|^{\alpha}, \tag{3.9}$$

for some finite positive constant $\gamma(I)$. For $\alpha = 1$, $k(s)$ is called locally Lipschitz continuous. Sub- and supersolution are defined analogously as before, so that, e.g., a subsolution $\underline{u}$ of (3.1), (3.8) is supposed to satisfy

$$B\underline{u} \leq g(x, \underline{u}).$$

The following basic result was then proven by Amann (1976c):

Theorem 3.1 Let D be a finite domain in E^N, $N \geq 2$, bounded by a $C^{2+\alpha}$-surface ∂D, with $\alpha \in (0, 1)$, such that D lies locally on one side of ∂D. Suppose that the following smoothness properties are satisfied:

(i) The coefficients of L satisfy

$$a^{ik} \in C^{2+\alpha}(D), \qquad b^l \in C^{1+\alpha}(D),$$

and

$$v \in C^{2+\alpha}(\partial D), \qquad h \in C^{1+\alpha}(\partial D).$$

(ii) f is α-Hölder continuous in its first variable and locally Lipschitz continuous in the second variable; g is locally Lipschitz continuous in both variables.

(iii) There exist a subsolution $\underline{u}$ and a supersolution $\bar{u}$, both in $C^2(D) \cap C^1(\partial D)$ and $\underline{u} \leq \bar{u}$.

Then there is at least one solution u of Eqs. (3.1) and (3.8) and $\underline{u} \leq u \leq \overline{u}$. Furthermore, there exists a minimal solution u_* and a maximal solution u^*, in the sense that for any solution u of (3.1), (3.8) one has

$$\underline{u} \leq u_* \leq u \leq u^* \leq \overline{u}.$$

The proof of Theorem 3.1 is given in Amann (1976c); it relies on functional-analytic methods.

Remarks The method of sub- and supersolutions does not always work as the following example illustrates. Consider the problem

$$Lu + f(u) = 0 \qquad \text{in} \quad D, \tag{3.10}$$

$$\frac{\partial u}{\partial N} := a^{ij}u_{,i}n_j = 0 \qquad \text{on} \quad \partial D, \tag{3.11}$$

$(n_j) = \mathbf{n} =$ outward normal on ∂D. Suppose that $f(0) = 0$ and $f(s) < 0$ for $s < 0$, $f(s) > 0$ for $s > 0$, and $df/ds \geq 0$ on $\mathbb{R}$. Suppose then $\underline{u}$ and $\overline{u}$ are sub- and supersolutions, respectively. Then, for $w = \overline{u} - \underline{u}$ we find that

$$Lw = -f(\overline{u}) + f(\underline{u}) \leq 0$$

since by assumption $\overline{u} \geq \underline{u}$. Because a nonconstant w must attain its minimum on ∂D, we check $\partial w/\partial N$ on ∂D. But,

$$\frac{\partial w}{\partial N} = \frac{\partial \overline{u}}{\partial N} - \frac{\partial \underline{u}}{\partial N} = 0,$$

which leads to a contradiction. Hence, w must be a constant and $Lw \equiv 0$, i.e., $f(\overline{u}) = f(\underline{u})$, implying that $\overline{u} = \underline{u}$. In other words, $\underline{u}$ and $\overline{u}$ must already be solutions. Another example of a case in which this method is not applicable is given by Kazdan and Warner (1975).

Let us now look at some examples where the method does work in order to illustrate its usefulness. Consider the nonlinear eigenvalue problem

$$\Delta u + \lambda f(u) = 0 \qquad \text{in} \quad D, \tag{3.12}$$

$$u = 0 \qquad \text{on} \quad \partial D, \tag{3.13}$$

where λ is a positive parameter, $f(0) > 0$ and f is increasing for positive argument. Clearly, $\underline{u} \equiv 0$ is an allowable subsolution so that it suffices to find a supersolution. The next result is an immediate consequence of Theorem 3.1.

Corollary 3.1 Let $|D|$ be the diameter of $D \subset E^n$. Then problem (3.12). (3.13) has a positive solution for any $\lambda \in (0, \lambda^*)$, and

$$\lambda^* \geq \frac{8n}{|D|^2} \sup_{t>0} \frac{t}{f(t)}.$$

Proof We choose the origin of our coordinate system in D such that D is contained in the ball

$$B = \{x \mid |x|^2 \leq \tfrac{1}{4}|D|^2\}.$$

We then take

$$\overline{u} = \alpha(\tfrac{1}{4}|D|^2 - |x|^2), \qquad \alpha > 0,$$

as a supersolution. It is not hard to see that this is an allowable choice as long as

$$\lambda f(\alpha \tfrac{1}{4}|D|^2) \leq 2n\alpha.$$

Setting $t := \alpha\frac{1}{4}|D|^2$, we see that this inequality is satisfied, provided that

$$\lambda \leq \frac{8n}{|D|^2} \sup_{t>0} \frac{t}{f(t)}.$$

The constant α can then be chosen appropriately.

Remarks (a) If $\lim_{t\to\infty} (f(t)/t) = 0$ it thus follows that problem (3.12), (3.13) has a positive solution for arbitrary $\lambda > 0$. On the other hand, Hudjaev (1964) has shown that a necessary condition for (3.12), (3.13) to have a positive solution for any $\lambda > 0$ is

$$\liminf_{t\to\infty} \frac{f(t)}{t} = 0.$$

Hudjaev also proved that a necessary and sufficient condition for Eqs. (3.12), (3.13) to be *unsolvable* for some $\lambda > 0$ is

$$\inf_{t>0} \frac{f(t)}{t} > 0.$$

Compare also with inequality (3.16).

(b) Sharper bounds for λ^* can be derived by taking $\overline{u} = \alpha\psi$ where

$$\Delta\psi + 1 = 0 \qquad \text{in} \quad D, \tag{3.14}$$

$$\psi = 0 \qquad \text{on} \quad \partial D. \tag{3.15}$$

Arguments similar to those given previously then show that the critical value λ^* in (3.12), (3.13) is bounded below by

$$\frac{1}{\psi_M} \sup_{t>0} \frac{t}{f(t)},$$

where $\psi_M = \max_D \psi(x)$.

(c) An optimal bound for λ^*, under the assumption that $f(0) > 0$ and f is increasing and convex, was given by Bandle and Hersch (1975). They showed that for a plane domain of given area the circle yields the minimum value of λ^*. A lower bound for λ^* that is optimal in a different sense will be derived in Chapter 6 (Theorem 6.1).

(d) Let ϕ be the (positive) first eigenfunction of

$$\Delta\phi + \lambda\phi = 0 \quad \text{in} \quad D, \qquad \phi = 0 \quad \text{on} \quad \partial D,$$

with corresponding eigenvalue $\lambda_1 > 0$. Clearly,

$$\int_D (\Delta\phi + \lambda_1\phi)u\,dx = 0$$

for any solution u of Eqs. (3.12) and (3.13). By Green's identity it follows that

$$\int_D (-\lambda f(u) + \lambda_1 u)\phi\,dx = 0,$$

and thus

$$\sup_D (-\lambda f(u) + \lambda_1 u) > 0,$$

and

$$\inf_D (-\lambda f(u) + \lambda_1(u) < 0.$$

In particular it follows that

$$\lambda_1 \inf_{s>0} \frac{s}{f(s)} < \lambda < \lambda_1 \sup_{s>0} \frac{s}{f(s)} \tag{3.16}$$

must hold. This last inequality implies that

$$\lambda^* \le \lambda_1 \sup_{s>0} \frac{s}{f(s)}.$$

(e) The existence of solutions in (3.12), (3.13) may also depend on the number of dimensions as the following arguments show. Rellich (1940) derived the following identity. For any sufficiently smooth function v, one has

$$\int_D \Delta v\, x^i v_{,i}\,dx = \frac{N-2}{2}\int_D |\nabla v|^2\,dx + \oint_{\partial D} \frac{\partial v}{\partial n} x^i v_{,i}\,ds - \frac{1}{2}\oint_{\partial D} x^i n_i |\nabla v|^2\,ds \tag{3.17}$$

Here N is the number of dimensions and $n = (n_i)$ is the outward normal on ∂D. For a solution u of (3.12), (3.13), (3.17) takes the form

$$\lambda N \int_D F(u)\,dx = \frac{N-2}{2}\int_D |\nabla u|^2\,dx + \frac{1}{2}\oint_{\partial D} b(s)\left(\frac{\partial u}{\partial n}\right)^2 ds, \tag{3.18}$$

with

$$F(u) := \int_0^u f(y)\,dy \qquad \text{and} \qquad b := x^i n_i.$$

If the domain D is "starshaped" with respect to some origin, i.e., $b(s) \geq 0$, then since

$$\int_D |\nabla u|^2\,dx = \lambda \int_D uf(u)\,dx,$$

we see that we must have

$$\int_D \left(NF(u) - \frac{N-2}{2}\, uf(u) \right) dx \geq 0.$$

In other words, there is *no* positive solution of (3.12), (3.13) if for $s > 0$, we have

$$F(s) < \frac{N-2}{2N}\, sf(s).$$

For example, if $f(s) = s^p$, there is no positive solution if

$$p > \frac{N+2}{N-2}. \tag{3.19}$$

Another interesting case is that in which $f(u) = e^u$ and $N = 2$ (see Bandle, 1975). In this case, the nonexistence result can be made sharper. Assume that $b(s) > 0$ and define

$$B := \oint_{\partial D} \frac{ds}{b(s)}.$$

Then, by Schwarz's inequality we have

$$B \oint_{\partial D} b(s) \left(\frac{\partial u}{\partial n} \right)^2 ds \geq \left(\oint_{\partial D} \frac{\partial u}{\partial n}\, ds \right)^2 = \lambda^2 \left(\int_D e^u\,dx \right)^2 =: \lambda^2 E^2,$$

and together with (3.18) we see that

$$2\lambda(E - A) \geq \tfrac{1}{2}\lambda^2 E^2 B^{-1},$$

or by completing squares

$$(\lambda E - 2B)^2 \leq 4B^2 - 4\lambda BA.$$

But the last inequality implies that

$$\lambda \leq \lambda^* \leq B/A \tag{3.20}$$

must hold. When ∂D is a circle, one has $B = 2\pi$, and it can be shown (see Bandle, 1975) that the equality sign holds in (3.20).

Another simple consequence of Theorem 3.1 is

Corollary 3.2 Assume that in problem (3.12), (3.13) $f(0) > 0$ and

$$\sup_{s>0} \frac{s}{f(s)} < \infty.$$

Then the critical value λ^* depends monotonically on the domain D.

Proof The proof is obvious: Let $\tilde{D}$ be a subdomain of D, and $\tilde{u}$ the solution of (3.12), (3.13) for $\tilde{D}$. As a subsolution in $\tilde{D}$ we choose $\underline{u} \equiv 0$ and as a supersolution we take $\bar{\tilde{u}} = u =$ solution for D. This works for any $\lambda \in (0, \lambda^*)$, and hence we have

$$\lambda^*_{\tilde{D}} \geq \lambda^*_{D}.$$

Note that since

$$\sup_{s>0} \frac{s}{f(s)} < \infty$$

we know by Remark (d) that $\lambda^* < \infty$. Finally, $u \equiv 0$ is *not* a solution of (3.12), (3.13) because of our assumption that $f(0) > 0$.

In many applications the boundary conditions are also nonlinear so that instead of (3.2) we have the boundary condition

$$Bu = g(x, u) \qquad \text{on} \quad \partial D. \tag{3.21}$$

The extension of Theorem 3.1 to the case of the boundary condition is also due to Amann (1976c):

Theorem 3.2 Assume that in addition to hypotheses (i)–(iii) of Theorem 3.1 $g(x, u)$ is locally Lipschitz continuous in its two arguments. Then the statement of Theorem 3.1 also holds for problem (3.1), (3.21).

The proof is given in Amann (1976c).

Of course, sub- and supersolutions are defined as before, so that, e.g., a supersolution $\bar{u}$ of (3.1), (3.2) has to satisfy $B\bar{u} \geq g(x, \bar{u})$ on ∂D.

As an application of Theorem 3.2, we consider the problem

$$Lu - a(x)u = 0 \qquad \text{in} \quad D, \tag{3.22}$$

$$\partial u/\partial \nu = g(x, u) \qquad \text{on} \quad \partial D, \tag{3.23}$$

as studied by Cohen (1971). Here $a(x) > 0$ in D is assumed to hold. The next result is a simple consequence of Theorem 3.2.

Corollary 3.3 Suppose g is Lipschitz continuous in both arguments, $g(x, 0) > 0$, and $g(x, u)$ is decreasing in u with $g(x, 1) = 0$. Then problem (3.22), (3.23) has a positive solution.

Proof We choose $\underline{u} \equiv 0$ and $\bar{u} \equiv 1$. It is obvious that these are allowable sub- and supersolutions. Furthermore, the trivial solution is excluded since $g(x, 0) > 0$.

Remarks (a) In his article, Cohen (1971) had to make the additional assumption that $g_{uu}(x, u) < 0$ for positive u and all $x \in D$.

(b) There is yet another extension of Theorem 3.1 to the case in which the nonlinear term also includes first-order derivatives of the solution u. With the calculations of Section 5.1 it is not hard to see that even in the one-dimensional case some growth restrictions have to be imposed if $f = f(x, u, \nabla u)$. Let f be Hölder continuous in its last two arguments and assume that for $f(x, \xi, \boldsymbol{\eta})$ the derivatives $\partial f/\partial \xi$ and $\partial f/\partial \eta^i$ ($\boldsymbol{\eta} = (\eta^i)$) are continuous on $D \times \mathbb{R}^{N+1}$ for $D \subset E^N$. Moreover, it is supposed that there exists a positive function $c(y)$, such that

$$|f(x, \xi, \boldsymbol{\eta})| \leq c(y)(1 + |\boldsymbol{\eta}|^2)$$

for every $y > 0$ and $(x, \xi, \boldsymbol{\eta}) \in D \times [-y, y] \times \mathbb{R}^N$. Under these assumptions on f and the remaining assumptions of Theorem 3.1, it was shown by Amann (1976a) that Theorem 3.1 can be extended to solutions of

$$Lu + f(x, u, \nabla u) = 0 \quad \text{in} \quad D, \tag{3.24}$$

$$Bu = g(x) \quad \text{on} \quad \partial D. \tag{3.25}$$

(c) Of course there are numerous other methods for proving existence in nonlinear elliptic problems, such as successive approximation or variational techniques. For representations of these topics the reader is referred to the books of, e.g., Miranda (1970), Gilbarg and Trudinger (1977), or Ladishenskaya and Ural'ceva (1968).

3.1.3 Existence Results in Nonlinear Parabolic Problems

Any solution of problem (3.1), (3.2) or (3.1), (3.21) can be considered as a possible steady state of a corresponding nonlinear parabolic problem. Indeed it is usually in this context that one will encounter nonlinear elliptic problems.

We shall therefore consider the following nonlinear initial-boundary value problem:

$$u_{,t} = Lu + f(t, x, u) \quad \text{in} \quad D \times (0, T), \tag{3.26}$$

$$Bu = g(t, x, u) \quad \text{on} \quad \partial D \times (0, T), \tag{3.27}$$

$$u(x, 0) = u_0(x). \tag{3.28}$$

Here L is a uniformly elliptic operator for $t \in (0, T)$, i.e.,

$$Lu = a^{ij}(x, t)u_{,ij} + b^i(x, t)u_{,i},$$

where the usual ellipticity requirement for the a^{ij} is supposed to hold for all $t \in (0, T)$ and a^{ij} and b^i are bounded functions of their arguments.

Sub- and supersolutions of (3.26)–(3.28) are now defined analogously as before: For a subsolution $\underline{u}$ the equality signs in (3.26)–(3.28) are to be replaced by the sign "$\leq$," and for a supersolution $\bar{u}$ by the sign "$\geq$."

If f and g are Lipschitz continuous in u, the maximum principle arguments of Section 3.12 again would show that if there exists a solution u and sub- and supersolution of (3.26)–(3.28), then we must have $\underline{u} \leq u \leq \bar{u}$. It is important that again the converse is true. More precisely one has

Theorem 3.3 (Pao, 1978; Czischke, 1978) Suppose the following conditions are satisfied:

(i) D, L, and B satisfy the smoothness requirements of Theorem 3.1.

(ii) g satisfies a Lipschitz condition in the argument u for any $(x, t) \in \bar{D} \times (0, T)$, and for any $y \geq 0$ there exists a $c(y) \geq 0$ such that

$$|f(t_2, x_2, u_2) - f(t_1, x_1, u_1)| \leq c(y)\{(|t_2 - t_1|^{1/2} + |x_2 - x_1|^2)^\alpha + |u_2 - u_1|\}$$

for $t_1, t_2 \in [0, T]$, $x_1, x_2 \in D$, $u_1, u_2 \in [-y, y]$.

(iii) There exist a subsolution $\underline{u}$ and a supersolution $\bar{u}$ of (3.26)–(3.28) such that

$$\underline{u} \leq \bar{u} \qquad \text{in} \quad D \times (0, T).$$

Then there exists a solution $\underline{u}$ of (3.26)–(3.28) and

$$\underline{u} \leq u \leq \bar{u} \qquad \text{in} \quad D \times (0, T).$$

A proof of Theorem 3.3 in the case where $\underline{u} \equiv 0$ is possible was first given by Pao (1978), and in the slightly more general version given, a different proof was presented in the dissertation of Czischke (1978).

Let us illustrate applications of Theorem 3.3 in some simple examples. A first observation may be stated as

Corollary 3.4 Assume that D, L, and f satisfy the smoothness requirements of Theorems 3.1 and 3.3, respectively, and

(i) For $(x, t) \in D \times (0, \infty)$ one has $0 \leq f(u, x, t) \leq h(u)$,

(ii) $0 \leq u_0(x) \leq M_0$ and $\int_{M_0}^\infty dy/h(y) = T_*$. Then problem (3.26), (3.28) with (3.27) replaced by Dirichlet boundary conditions has a positive solution for any $t \in (0, T)$ and $T \geq T_*$.

Proof With our assumptions on f and u_0, $\underline{u} \equiv 0$ is obviously an allowable subsolution. As a supersolution $\bar{u}$ we select the solution $z(t)$ of the ordinary differential equation

$$\dot{z} = h(z), \qquad z(0) = M_0.$$

It is easily checked that $z(t)$ is indeed a supersolution which remains finite for any $0 \le t < T_*$, where

$$T_* = \int_{M_0}^{\infty} \frac{dy}{h(y)}.$$

This proves Corollary 3.4.

Note that it follows in particular that if

$$\int_{M_0}^{\infty} \frac{dy}{h(y)} = \infty,$$

then the solution u exists for all time.

In population dynamics the following problem is of interest:

$$u_{,t} = \Delta u + f(u) \qquad \text{in} \quad \Omega \times (0, \infty), \tag{3.29}$$

$$\frac{\partial u}{\partial n} = 0 \qquad \text{on} \quad \partial\Omega \times (0, \infty), \tag{3.30}$$

$$u(x, 0) = u_0(x) > 0. \tag{3.31}$$

In this context it is often assumed that $f(0) = 0$ and $f(u) > 0$ for $0 < u < 1$, and $f(1) = 0$ and $f(u) < 0$ for $u > 1$.

The next result is again an immediate consequence of Theorem 3.3.

Corollary 3.5 Under the previous assumptions on $f(u)$ problem (3.29)–(3.31) has a bounded solution for all time.

Proof Again we may take $\underline{u} = 0$ and $u = z(t)$, where

$$\dot{z} = f(z), \qquad z(0) = \max_D u_0(x) < \infty.$$

Clearly, $\lim_{t\to\infty} z(t) = 1$, implying that the solution u of (3.29)–(3.31) will stay bounded for all time.

In problem (3.29)–(3.31), we only have the two possible steady states: $u = 0$ or $u \equiv 1$. No nonconstant steady state is possible, as another simple maximum principle argument would readily show.

Let us suppose now that $0 < u_0(x) < \infty$ and $u_0(x) \not\equiv 0$ holds. It is easy to see that any nonzero initial distribution will lead to the steady state $u \equiv 1$. To prove this simple fact, assume contrariwise that

$$\lim_{t\to\infty} u(x, t) \equiv 0$$

Then there must exist a T such that

$$0 \le u(x, t) < 1 \qquad \text{for} \quad t > T.$$

Consider now the average value $a(t)$, defined as

$$a(t) := \int_D u(x,t)\,dx.$$

Then

$$\dot{a}(t) = \int_D u_{,t}\,dx = \int_D f(u(x,t))\,dx > 0$$

since u has vanishing normal derivative on ∂D. Therefore $a(t)$ must be increasing for $t > T$, which in turn clearly contradicts the assumption that $\lim_{t\to\infty} u(x,t) \equiv 0$.

Hence, we may summarize our result by saying that for any nonzero distribution u_0 in problem (3.29)–(3.31), the solution will exist for all time and tend to the uniform steady state $u \equiv 1$.

As a last example let us consider the following absorption problem:

$$u_{,t} = \Delta u \qquad \text{in} \quad D \times (0,\infty) \tag{3.32}$$

$$\partial u/\partial n = g(u) \qquad \text{on} \quad \partial D \times (0,\infty) \tag{3.33}$$

$$u(x,0) = u_0(x) > 0. \tag{3.34}$$

We assume that $g(0) > 0$, $g'(u) \leq 0$, and $g(u) \geq 0$ for $u > 0$. Under these assumptions on g it is easy to prove

Corollary 3.6 Problem (3.32)–(3.34) has a positive solution for any $t \geq 0$.

Proof Because of our assumptions we may choose $\underline{u} \equiv 0$. As a supersolution we select

$$\bar{u}(x,t) = v(t + \psi(x)) =: v(s),$$

where v and ψ are yet to be determined. Now $\bar{u}$ satisfies

$$\bar{u}_{,t} - \Delta\bar{u} = \frac{dv}{ds}(1 - \Delta\psi) - \frac{d^2v}{ds^2}|\nabla\psi|^2,$$

and

$$\frac{\partial\bar{u}}{\partial n} - g(\bar{u}) = \frac{\partial\psi}{\partial n}\frac{\partial v}{ds} - g(v).$$

A convenient way of choosing $v(z)$ and $\psi(x)$ is therefore as follows. We let ψ be the solution of

$$\Delta\psi = 1 \quad \text{in} \quad D, \qquad \frac{\partial\psi}{\partial n} = \beta = \text{const} \quad \text{on} \quad \partial D,$$

where ψ is normalized such that

$$\min_D \psi(x) = \max_D u_0(x) = M_0.$$

Note that it follows from Green's identity that

$$\beta = \frac{V(D)}{S(\partial D)}, \qquad V(D) = \text{volume of } D, \qquad S(\partial D) = \text{surface area of } \partial D.$$

We now define $v(s)$ to be the solution of

$$\frac{dv}{ds} = \frac{1}{\beta} g(v), \qquad v(M_0) = M_0.$$

It is easily checked that with our assumptions on $g(v)$, $v(s)$ exists for all $S > M_0$ and is bounded for finite s. In addition, we have

$$\frac{dv}{ds} \geq 0, \qquad \frac{d^2v}{ds^2} \leq 0,$$

so that $\bar{u} = v(t + \psi(x))$ satisfies

$$\bar{u}_{,t} - \Delta \bar{u} \geq 0 \qquad \text{in} \quad D \times (0, \infty),$$

$$\frac{\partial \bar{u}}{\partial n} \geq g(\bar{u}) \qquad \text{on} \quad \partial D \times (0, \infty),$$

and

$$\bar{u}(x, 0) = v(\psi(x)) \geq v(M_0) = M_0 \geq u_0(x).$$

Hence, $\bar{u} = v(t + \psi(x))$ is a positive supersolution, which by Theorem 3.3 implies that problem (3.32)–(3.34) has a positive solution.

Remarks (a) Corollary 3.6 is a special case of a result of Walter (1975). He considered the more general case in which the Laplacian is replaced by a uniformly elliptic operator, and the equality signs in (3.32), (3.33) are replaced by inequality signs. He then showed that if

$$\int^{\infty} \frac{ds}{g(s)g'(s)}$$

converges, the solution of (3.32)–(3.34) must blow up in finite time, but if the integral diverges one has the case of global existence. A different blow-up result in problem (3.32)–(3.34) will be proven in Corollary 9.8.

(b) A very general existence result for a system of parabolic equations was proven by Amann (1978). He considered a system of the form

$$\begin{aligned} u^i_{,t} &= a^i_{jk}(x,t)u^i_{,jk} + b^i_j(x,t)u_{,j} \\ &\quad + a^i(x,t)u + f^i(x,t,u^j,\nabla u^j) \qquad \text{in} \quad D \times (0,T), \end{aligned}$$

$$Bu^i = 0 \qquad \text{on} \quad \partial D \times (0,T), \qquad i, j = 1, \ldots, m.$$

An essential feature of such systems, containing first derivatives of the unknown functions, is again that a growth restriction has to be imposed, similar to the one required in problem (3.24), (3.25). For more details the reader is referred to the original article of Amann (1978).

3.2 SOME REGULARITY RESULTS

3.2.1 Elliptic Problems

The way the maximum principles will be applied in later chapters requires in most cases that the solutions of various problems have higher derivatives than those that appear in their respective differential equations. So, e.g., we shall consider a nonlinear Dirichlet problem, the solutions of which are required to have derivatives of order three instead of only two. At first sight this may seem to be a restrictive assumption. The purpose of this section is therefore to review some of the known results on the regularity of the solutions of elliptic problems. It is probably one of the most important intrinsic properties of solutions of second-order elliptic problems that in many instances they have derivatives of order higher than two.

Let us start with a classical example. Let h be a harmonic function in some finite domain D of E^N, i.e.,

$$\Delta h = 0 \qquad \text{in} \quad D. \tag{3.35}$$

Thus it is only assumed that $h \in C^2(D)$. However, as a consequence of the well-known Poisson integral representation of h in the interior of D, it follows that h is even analytic in D (see Courant and Hilbert, 1953).

In his famous paper, Bernstein (1904) proved that the same analyticity holds also for solutions of

$$\Delta u + \lambda u = 0 \qquad \text{in} \quad D \in E^2. \tag{3.36}$$

Bernstein showed that if $u \in C^2(D)$, then u is even analytic in D. Since his proof was based on complex variable arguments, the result could only be proven for plane domains. An important consequence of the analyticity of the solution is the so-called strong unique continuation property: If u (or h) has a zero of infinite order at some interior point of D, it follows that u must vanish identically in D. The strong unique continuation property holds for much larger classes of problems as well, as we shall see later on.

As a next step, one may consider solutions of a linear elliptic equation of the form

$$a^{ij}(x)u_{,ij} + b^i(x)u_{,i} + c(x)u = 0 \qquad \text{in} \quad D \in E^N. \tag{3.37}$$

It is quite obvious that if we want the solution to be very smooth, we must also have smooth coefficients a^{ij}, b^i, c.

The generalization of Berstein's classical result on the analyticity of solutions of (3.36) to solutions of (3.37) is essentially due to Hadamard (1932). He showed that if the coefficients in (3.37) are analytic, there exists a fundamental solution of (3.37), which implies that the solutions of (3.37) are analytic.

An important date in the study of nonlinear elliptic equations is probably the year 1900. In that year at the International Congress of Paris, Hilbert stated his conjecture that every solution of an analytic elliptic equation is analytic. Here u is a solution of

$$F(u_{,ik}, u_{,i}, u, x) = 0 \qquad \text{in} \quad D \subset E^N, \tag{3.38}$$

where F is an analytic function of all its arguments and the quadratic form $Q(\xi, \xi)$, *defined as*

$$Q(\xi, \xi) = \frac{\partial F}{\partial u_{,ik}} \xi^i \xi^k,$$

is positive definite for all values of the arguments.

For plane domains, Hilbert's conjecture was proved by Bernstein (1926). A fundamental idea that Berstein employed in his proof was that the existence of analytic solutions is ensured once certain a priori bounds for the solutions can be established.

If the function F is not analytic in all its arguments, but still possesses derivatives up to a certain order, one will expect that the solutions of (3.38) have derivatives of order higher than two. Such a result was proven by Hopf (1931), and we shall state it as

Theorem 3.4 Let F be continuous with all its derivatives up to order $n\ (\geq 1)$ and suppose the nth derivatives of F with respect to $u_{,ik}$ are Lipschitz continuous and α-Hölder continuous with respect to the remaining variables. Then every solution u of (3.38) of class $C^{2+\alpha}(D)$ is also of class $C^{n+2+\alpha}(D)$.

The result of Hopf was subsequently strengthened and generalized by different people. For a complete bibliography on this subject the reader is referred to the book of Miranda (1970).

The extension of Hopf's result may be stated as

Theorem 3.5 (i) Suppose that the assumptions of Theorem 3.4 on F are satisfied and u is a $C^2(D)$-solution of (3.38). Then u is also of class $C^{n+2+\alpha}(D)$.

(ii) Let Γ be an open portion of ∂D. If ∂D is of class $C^{n+2+\alpha}$ and u is of class $C^{n+2+\alpha}$ on Γ and of class C^2 in $(D - \partial D) \cap \Gamma$, then u is also of class $C^{n+2+\alpha}$ in $(D - \partial D) \cap \Gamma$.

It is important to note that, roughly speaking, if the the solution of an elliptic equation is to be smooth on the boundary ∂D of the given domain D, then the boundary must have the same smoothness properties. This is in contrast to the smoothness of u in the interior of D, where no assumptions on the regularity of ∂D have to be made a priori.

As we have seen earlier, the strong unique continuation property is in some sense a substitute for analyticity.

For a list of references on the next two theorems of this section the reader may again consult the book of Miranda (1970). We state the next result as

Theorem 3.6 Suppose that F and $\partial F/\partial u_{,ik}$ are Lipschitz continuous and $F(0,0,0,x) = 0$. Then for $C^2(D)$-solutions of (3.38) the strong unique continuation property holds.

An important special class of equations are the so-called quasi-linear elliptic equations. They are of the form

$$a^{ik}(x, u, u_{,j})u_{,ik} = f(x, u, u_{,j}). \tag{3.39}$$

Suppose that the following growth restriction for f holds:

$$f(x, u, u_{,j}) = O(|u| + \sqrt{u_{,j}u_{,j}}). \tag{3.40}$$

Then the following theorem holds:

Theorem 3.7 Suppose that the a_{ik} are Lipschitz continuous and (3.40) holds. Then for $C^2(D)$-solutions of (3.39) the strong unique continuation property holds.

The regularity results cited in this section can now be combined with the existence results of Section 3.1.2. For example, the combination of Corollary 3.1 and Theorem 3.5 leads to

Corollary 3.7 Let D be a finite domain in E^N bounded by a $C^{2+\alpha}$ surface ∂D with $\alpha \in (0,1)$. Suppose that $f(s) \in C^{n+\alpha}(I)$ for any finite subinterval of $\mathbb{R}^+$, and $f(0) > 0$ and $\sup_{s>0}(s/f(s)) < \infty$. Then the problem

$$\Delta u + \lambda f(u) = 0 \quad \text{in} \quad D, \qquad u = 0 \quad \text{on} \quad \partial D \tag{3.41}$$

has a positive solution of class $C^{n+2+\alpha}$ in D for any $\lambda \in (0, \lambda^*)$ and

$$\lambda^* \geq \frac{8N}{|D|^2} \sup_{s>0} \frac{s}{f(s)}.$$

It is quite clear that Corollary 3.7 can easily be generalized to solutions of more general elliptic equations such as the cases in which the Laplacian in Eq. (3.41) is replaced by a uniformly elliptic operator L, or $f(u)$ by $f(u, x)$ under the appropriate assumptions on L and f.

In conclusion let us mention that Hilbert's conjecture can finally be answered affirmatively (see Miranda, 1970: Theorem 44, IV) based on a number of rather deep results on the regularity properties of solutions of (3.38).

3.2.2 Parabolic Problems

A solution u of

$$\Delta u - u_{,t} = 0 \qquad \text{in} \quad D \subset E^N \times \mathbb{R}^+ \tag{3.42}$$

can be considered as the parabolic counterpart of a harmonic function. However, since the time t appears only in a first-derivative term in (3.42), it is quite clear that we cannot expect the same regularity behavior of u with respect to spatial differentiation and differentiation with respect to t.

It is a well-known fact that any classical solution u of (3.42) in D is analytic in the spatial variables, and C^∞ in t, but in general, not analytic in t.

Let us next consider a more general linear parabolic equation of the form

$$u_{,t} - a^{ij}(x,t)u_{,ij} - b^i(x,t)u_{,i} - c(x,t)u = f(x,t) \qquad \text{in} \quad D \subset E^N \times \mathbb{R}^+. \tag{3.43}$$

The following three theorems are all proven in the book of Friedman (1964). The first result describes how the smoothness of the coefficients a^{ij}, b^i, c, and f implies the smoothness of the solutions. In fact one has

Theorem 3.8 Let u be a classical solution of (3.43) and suppose that a^{ij}, b^i, c, and f are of class $C^{m+\alpha}$ in D with respect to x. Then u is also of class $C^{m+2+\alpha}$ in D and $u_{,t}$ is of class $C^{m+\alpha}$ with respect to x.

In the preceding theorem only smoothness in the spatial variables has been considered. It is also of interest to investigate differentiability with respect to the time t.

Let us denote a general kth derivative with respect to t by D_t^k, and a mth derivative with respect to the spatial variables by D_x^m. Then the following result holds.

Theorem 3.9 Suppose u is a classical solution of Eq. (3.43) and for a^{ij}, b^i, c, and f $D_x^m D_t^k$ exist for $0 \le m + 2k \le p$, $k \le q$, and are α-Hölder continuous in the domain D. Then for $0 \le m + 2k \le p + 2$, $k \le q + 1$, $D_x^m D_t^k u$ exist and are also α-Hölder continuous.

Let $\partial/\partial v$ denote the conormal derivative with respect to a^{ij}. Then the following analyticity result is proven in Friedman (1969).

Theorem 3.10 Let u be a classical solution of Eq (3.43) with $D = \Omega \times (0, T)$, satisfying $\partial u/\partial v = 0$ on $\partial\Omega \times (0, T)$. If f and a^{ij}, b^i, c are analytic in $\bar{\Omega} \times [0, T]$, and $\partial\Omega$ is analytic, then u is analytic in $\bar{\Omega} \times (0, T)$.

Note that the analyticity of u with respect to t for $t = 0$ and $t = T$ is not ensured.

Finally, let us mention a regularity theorem that is valid for solutions of a general nonlinear parabolic equation of the form

$$F(x, t, u, u_{,i}, u_{,ij}, u_{,t}) = 0. \tag{3.44}$$

It is assumed that $\partial F/\partial u_{,t} < 0$ holds and the quadratic form

$$Q(\xi, \xi) = \frac{\partial F}{\partial u_{,ik}} \xi^i \xi^k$$

is positive definite for all values of the arguments.

As a last result we cite

Theorem 3.11 Let u be a classical solution of (3.44) and assume that $F \in C^{p+\alpha}(D \times \mathbb{R}^{N^2+N+1})$. If F is parabolic in D with respect to u and if $D_x u$, $D_x^2 u$, $D_t u$ are Hölder continuous with some exponent ε, then

$$D_x^m u, \qquad D_t D_x^k u, \qquad 0 \le m \le p+3, \quad 0 \le k \le p+1,$$

exist and are α-Hölder continuous in D.

The proof of Theorem 3.11 is given in Friedman (1964).

Remarks For the function $F(x, t, u, u_{,i}, u_{,ij}, u_{,t})$ the "distance" between two points $\xi = (x, t, \eta_k)$ and $\bar{\xi} = (\bar{x}, \bar{t}, \bar{\eta}_k) \in D \times \mathbb{R}^{N^2+N+1}$ is measured by

$$d(\xi, \bar{\xi}) := \left\{ |x - \bar{x}|^2 + |t - \bar{t}| + \sum_{k=1}^{N^2+N+1} (\bar{\eta}_k - \eta_k)^2 \right\}^{1/2}.$$

For the function $F(\xi)$ Hölder continuity is defined in the usual sense, i.e.,

$$|F(\xi) - F(\bar{\xi})| \le C(d(\xi, \bar{\xi}))^\alpha$$

is supposed to hold now for some finite constant C and all "points" $\xi, \bar{\xi} \in D \times \mathbb{R}^{N^2+N+1}$. For more details and other regularity results the reader is referred to the books of Friedman (1963, 1964, 1969).

Chapter 4
DIFFERENTIAL GEOMETRY

In this chapter some differential geometric formulas, which will be of importance later on, will be derived. A certain degree of familiarity with the concepts of differential geometry is assumed since the choice of topics presented in this chapter is very one-sided. The main purpose is to derive some relations connected with the Laplacian and gradient first. In the third section, some basic notions of Riemannian geometry have to be introduced for the future treatment of applications in the context of elliptic or parabolic problems.

In the following we shall frequently write $\mathbf{v}$ for a vector with components v^i. The summation convention will be used throughout this chapter also. Thus, for instance, the scalar product $\mathbf{v} \cdot \mathbf{w}$ will also be written as $v^i w^i$.

4.1 THE FRENET FORMULAS AND SOME APPLICATIONS

Let C be a curve in the plane given in parametric representation by

$$x^i = x^i(s), \qquad i = 1, 2,$$

where s denotes the arc length along C measured from some point on C. Then for the tangent vector $\boldsymbol{\tau}$ we have

$$\tau^i = \dot{x}^i, \qquad \tau^i \tau^i = |\boldsymbol{\tau}|^2 = 1, \qquad \cdot = \frac{d}{ds}.$$

By differentiation we find

$$\dot{x}^i\ddot{x}^i \equiv \dot{\mathbf{x}} \cdot \ddot{\mathbf{x}} = 0,$$

i.e., $\ddot{\mathbf{x}}$ must be parallel to the normal vector $\mathbf{n}$. Thus with some scalar k, the relation

$$\ddot{\mathbf{x}} = k\mathbf{n}$$

must hold. The scalar k is called the curvature of C at a given point of C. Since $|\mathbf{n}| = 1$, another differentiation with respect to s yields

$$n^i\dot{n}^i = 0,$$

i.e.,

$$\dot{n}^i = \alpha\tau^i$$

for some scalar α. Finally using that $n^i\tau^i = 0$, we see that

$$\dot{n}^i\tau^i + n^i\dot{\tau}^i = 0,$$

i.e., $\alpha = -k$. Combining our previous results we arrive at

$$\dot{\tau}^i = kn^i, \qquad \dot{n}^i = -k\tau^i. \tag{4.1}$$

This system is called the *Frenet formulas.*

Remarks (a) Note that the sign of the curvature is given once the parametric representation $\mathbf{x} = \mathbf{x}(s)$ has been chosen. Later on we shall almost always deal with closed simple curves, namely, the boundary of a plane domain D. Then the normal $\mathbf{n}$ on ∂D is usually the normal pointing outward. The sign of the curvature, however, is normally taken in such a way that the curvature is nonnegative if the domain is convex. This is just the opposite sign of what one would find from (4.1).

(b) Various formulas for the curvature of a plane curve are not hard to derive and can be found in almost any textbook on differential geometry.

Assume now that a function $u(x^i)$ is defined in some region R bounded on one side by curve C (see Figure 4.1). We define $\mathbf{n}$ as the unit normal vector pointing outward from R. Let x be a point on C and y a nearby point of R, so that the coordinates of x are $x^i(s)$ and those of y $x^i(s) - \varepsilon n^i(s)$. In the neighborhood of x we can define

$$u(x^i(s) - \varepsilon n^i(s)) =: u(s, \varepsilon).$$

We then define the "normal derivative" $\partial u/\partial n$ of u by

$$\frac{\partial u}{\partial n} = \lim_{\varepsilon \to 0} \frac{1}{\varepsilon}(u(x^i) - u(x^i - \varepsilon n^i)).$$

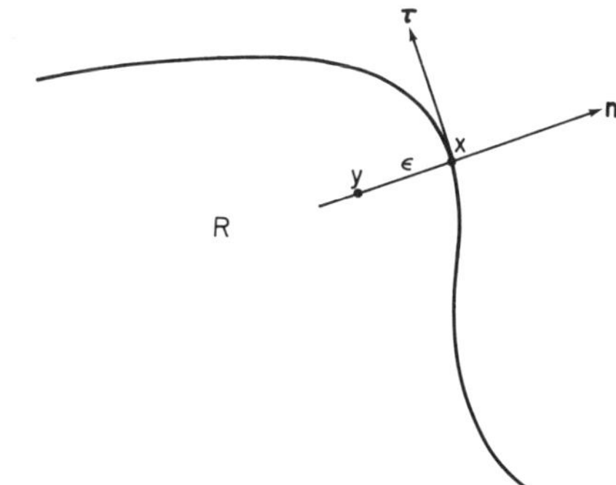

Figure 4.1

On C we have

$$u(x^i) = u(x^i(s)) = u(s),$$

so that we can also define a tangential derivative by

$$\frac{\partial u}{\partial s} = \frac{du}{ds} = \frac{\partial u}{\partial x^i}\dot{x}^i = u_{,i}\tau^i = \nabla u \cdot \boldsymbol{\tau} \qquad \text{(summation convention).}$$

On the other hand, we have

$$\partial u/\partial n = u_{,i}n^i = \nabla u \cdot \mathbf{n}.$$

We now want to define second derivatives. It seems logical to set

$$\frac{\partial^2 u}{\partial s^2} = \frac{d}{ds}(\nabla u(s) \cdot \boldsymbol{\tau}(s)) = \frac{d}{ds}(u_{,i}(s)\dot{x}^i(s))$$
$$= u_{,ij}\dot{x}^i\dot{x}^j + u_{,i}\ddot{x}^i(s) = u_{,ij}\tau^i\tau^j - ku_{,i}n^i.$$

Here, use of the Frenet formulas and our orientation of $\mathbf{n}$ have been made in the last step. We define next

$$\frac{\partial^2 u}{\partial n^2} := \frac{\partial^2}{\partial \varepsilon^2}(u(s,\varepsilon))\bigg|_{\varepsilon=0} = \frac{\partial}{\partial \varepsilon}(u_{,i}(s,\varepsilon)n^i(s))\bigg|_{\varepsilon=0} = u_{,ij}n^in^j.$$

Finally, mixed derivatives may be considered as well. We set

$$\frac{\partial}{\partial s}\left(\frac{\partial u}{\partial n}\right) = \frac{d}{ds}(u_{,i}(s)n^i(s)) = u_{,ij}n^i\tau^j + u_{,i}\dot{n}^i$$
$$= u_{,ij}n^i\tau^j + ku_{,i}\tau^i.$$

On the other hand, we may define

$$\frac{\partial}{\partial n}\left(\frac{\partial u}{\partial s}\right) \quad \text{by} \quad \frac{\partial}{\partial n}\left(\frac{\partial u}{\partial s}\right) := \frac{\partial}{\partial \varepsilon}(u_{,i}(s,\varepsilon)\tau^i(s)) = u_{,ij}\tau^in^j.$$

According to the previously given definitions, we have

$$\frac{\partial}{\partial n}\left(\frac{\partial u}{\partial s}\right) = \frac{\partial}{\partial s}\left(\frac{\partial u}{\partial n}\right) - k\frac{\partial u}{\partial s}. \tag{4.2}$$

It is then easy to check that a combination of the previous formulas yields

$$\frac{\partial^2 u}{\partial n^2} + k\frac{\partial u}{\partial n} + \frac{\partial^2 u}{\partial s^2} = u_{,ii} = \Delta u \tag{4.3}$$

on C. In (4.3) we have made use of the fact that $\tau^1 = -n^2, \tau^2 = n^1$, and $(\tau^1)^2 + (\tau^2)^2 = (n^1)^2 + (n^2)^2 = 1$. Formula (4.3) will be used numerous times in later chapters.

Example Let r, ϕ be the usual polar coordinates in the plane and the curve C be a circle of radius r. Then on C we have $s = r\phi$, so that relation (4.3) becomes

$$\frac{\partial^2 u}{\partial r^2} + \frac{1}{r}\frac{\partial u}{\partial r} + \frac{1}{r^2}\frac{\partial^2 u}{\partial \phi^2} = \Delta u.$$

4.2 SURFACES

Let S be a surface in space given in parametric representation by

$$x^i = x^i(y^\alpha), \qquad i = 1, 2, 3, \qquad \alpha = 1, 2.$$

It is assumed that the surface is smooth (say C^1) and that the vector product $\mathbf{x}_{,1} \times \mathbf{x}_{,2}$ does not vanish. Greek indices will always run from 1 to 2, and the summation convention is used for both kinds of indices as well.

The line element on S is given by

$$\begin{aligned} ds^2 &= |x_{,\alpha}\, dy^\alpha|^2 = \left|\frac{\partial \mathbf{x}}{\partial y^\alpha} \cdot dy^\alpha\right|^2 \\ &= x^i_{,\alpha}\, dy^\alpha x^i_{,\beta}\, dy^\beta = x^i_{,\alpha} x^i_{,\beta}\, dy^\alpha dy^\beta := g_{\alpha\beta}\, dy^\alpha\, dy^\beta. \end{aligned} \tag{4.4}$$

Here we have defined the covariant metric tensor $g_{\alpha\beta}$ by the last equation in (4.4).

The scalar product of two vectors $\mathbf{a} = a^\alpha$ and $\mathbf{b} = b^\beta$ in a tangent plane of S is given by

$$\mathbf{a} \cdot \mathbf{b} = \mathbf{x}_{,\alpha} a^\alpha \mathbf{x}_{,\beta} b^\beta = x^i_{,\alpha} x^i_{,\beta} = g_{\alpha\beta} a^\alpha b^\beta.$$

The area element dA of S is given by

$$dA = |\mathbf{x}_{,1}\, dy^1 \times \mathbf{x}_{,2}\, dy^2| = |(|\mathbf{x}_{,1}|^2 |\mathbf{x}_{,2}|^2 - (\mathbf{x}_{,1} \cdot \mathbf{x}_{,2})^2)^{1/2}\, dy^1 dy^2.$$

We can rewrite the expression above as

$$\begin{aligned} dA &= (x^i_{,1}x^i_{,1}x^j_{,2}x^j_{,2} - (x^i_{,1}x^i_{,2})^2)^{1/2}\,dy^1\,dy^2 \\ &= (g_{11}g_{22} - (g_{12})^2)^{1/2}\,dy^1\,dy^2 \\ &= (\det(g_{\alpha\beta}))^{1/2}\,dy^1\,dy^2. \end{aligned} \tag{4.5}$$

It is customary to denote the determinant of $g_{\alpha\beta}$ by g so that we have

$$dA = \sqrt{g}\,dy^1\,dy^2.$$

Example A sphere of radius R centered at the origin can be given in parametric representation as

$$\mathbf{x}(y^1, y^2) = R(\cos y^2 \cos y^1, \cos y^2 \sin y^1, \sin y^2)$$

(see Figure 4.2). The metric tensor $g_{\alpha\beta}$ is given in this case by

$$g_{\alpha\beta} = \begin{bmatrix} R^2\cos^2 y^2 & 0 \\ 0 & R^2 \end{bmatrix},$$

and therefore the area element is given by

$$dA = R^2 \cos^2 y\,dy^1\,dy^2.$$

A curve C on S can be given in the form

$$x^i = x^i(y^\alpha(t)),$$

where t is an arbitrary parameter. We can interpret C as the image of a curve $\tilde{C}$ given by

$$\tilde{C}: \quad y^\alpha = y^\alpha(t)$$

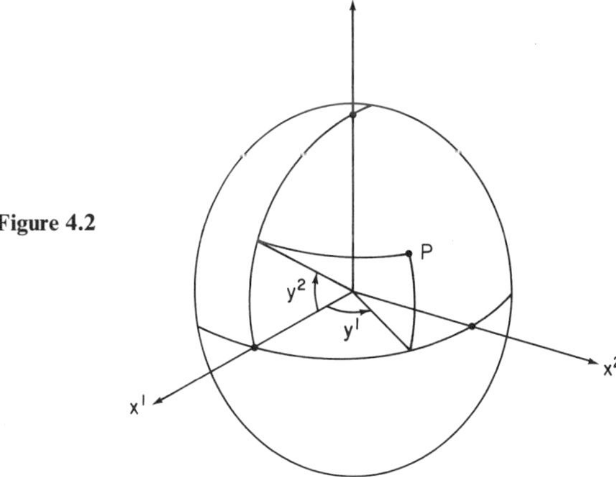

Figure 4.2

in the "parameter plane." In particular, we may choose the parameter t such that it measures the arc length along C on S. We then write s in the place of t and denote a derivative with respect to s by an overdot, whereas we shall use a prime for a derivative with respect to an arbitrary parameter t. The vector

$$\dot{\mathbf{x}} = x^i_{,\alpha}\dot{y}^\alpha$$

therefore lies in a tangent plane and

$$|\dot{\mathbf{x}}|^2 = x^i_{,\alpha}x^i_{,\beta}\dot{y}^\alpha\dot{y}^\beta = 1 = g_{\alpha\beta}\dot{y}^\alpha\dot{y}^\beta.$$

The vector

$$\ddot{\mathbf{x}} = (x^i_{,\alpha}\dot{y}^\alpha)^{\dot{}} = x^i_{,\alpha\beta}\dot{y}^\alpha\dot{y}^\beta + x^i_{,\alpha}\ddot{y}^\alpha$$

is called the curvature vector of C. The length of the component of $\ddot{\mathbf{x}}$ parallel to the normal n to S is called the "normal curvature" k_n. The length of the component of $\ddot{\mathbf{x}}$ lying in the tangent plane to S is called the "geodesic curvature" k_g. Thus by definition

$$k_n = \ddot{\mathbf{x}} \cdot \mathbf{n} = (x^i_{,\alpha\beta}\dot{y}^\alpha\dot{y}^\beta + x^i_{,\alpha}\ddot{y}^\alpha)n^i = x^i_{,\alpha\beta}n^i\dot{y}^\alpha\dot{y}^\beta =: h_{\alpha\beta}\dot{y}^\alpha\dot{y}^\beta$$

since $x^i_{,\alpha}n^i = 0$. The form $h_{\alpha\beta}\dot{y}^\alpha\dot{y}^\beta$ is called the second fundamental form of S, and $h_{\alpha\beta}$ the second (covariant) fundamental tensor. Note also that $g_{\alpha\beta} = g_{\beta\alpha}$ and $h_{\alpha\beta} = h_{\beta\alpha}$. The geodesic curvature of C can be calculated by the triple product

$$k_g = (\mathbf{n}, \dot{\mathbf{x}}, \ddot{\mathbf{x}}) = \det\begin{vmatrix} n^1 & \dot{x}^1 & \ddot{x}^1 \\ n^2 & \dot{x}^2 & \ddot{x}^2 \\ n^3 & \dot{x}^3 & \ddot{x}^3 \end{vmatrix}.$$

If t is an arbitrary parameter, then the vector

$$\frac{y^{\alpha'}}{(g_{\alpha\beta}y^{\alpha'}y')^{1/2}}$$

has unit length. Therefore we have

$$k_n = \frac{h_{\alpha\beta}y^{\alpha'}y^{\beta'}}{g_{\alpha\beta}y^{\alpha'}y^{\beta'}} = \frac{h_{\alpha\beta}\,dy^\alpha\,dy^\beta}{g_{\alpha\beta}\,dy^\alpha\,dy^\beta}. \tag{4.6}$$

Equation (4.6) shows that k_n depends only on the "direction" dy^α in the tangent plane and *not on* C. The normal curvature k_n is therefore also equal to the curvature of the normal section S in the direction dy^α. It is thus natural to ask for the maximum and minimum of k_n at a given point of S. It is well known that the maximum and minimum are given by the eigenvalues $\kappa = \kappa_1$ and $\kappa = \kappa_2$ of

$$h_{\alpha\beta} - \kappa g_{\alpha\beta}, \qquad \text{i.e.,} \quad \det(h_{\alpha\beta} - \kappa g_{\alpha\beta}) = 0.$$

The two invariants

$$2H := \kappa_1 + \kappa_2 \qquad \text{and} \qquad K_G := \kappa_1 \kappa_2$$

are called the "mean curvature" and the "Gaussian curvature" of S at a given point. For further development, it is useful to define the "contravariant metric tensor" $g^{\alpha\beta}$ by

$$g^{\alpha\beta} = (g_{\alpha\beta})^{-1} \qquad \text{(inverse tensor)} \tag{4.7}$$

and the mixed second fundamental tensor h_α^β by

$$h_\alpha^\beta := g^{\beta\gamma} h_{\alpha\gamma}. \tag{4.8}$$

It then follows that since $(h_{\alpha\gamma} - \kappa g_{\alpha\gamma}) g^{\beta\gamma} = h_\alpha^\beta - \kappa \delta_\alpha^\beta$, we have

$$2H = h_\alpha^\alpha = h_1^1 + h_2^2 \tag{4.9}$$

and

$$K_G = \det(h_\alpha^\beta). \tag{4.10}$$

The famous *Theorema Egregium* of Gauss states that the Gaussian curvature K_G depends only on the first fundamental tensor $g_{\alpha\beta}$ and its first and second derivative with respect to the y^α. For the proof of this result the reader may consult any of the standard texts in differential geometry.

For later applications we shall also need the derivatives $n^i_{,\alpha}$ of the unit normal vector. Now since $|\mathbf{n}| = 1$, we find by differentiation that

$$n^i n^i_{,\alpha} = 0.$$

Therefore, with some scalars c_α^β we must have

$$n^i_{,\alpha} = c_\alpha^\beta x^i_{,\beta}.$$

Furthermore, since $\mathbf{x}_{,\alpha} \cdot \mathbf{n} = x^i_{,\alpha} n^i = 0$, it follows by differentiation that

$$x^i_{,\alpha\beta} n^i + x^i_{,\alpha} n^i_{,\beta} = 0.$$

Combining our two last relations and using the fact that by definition $x^i_{,\alpha\beta} n^i = h_{\alpha\beta}$, we arrive at

$$h_{\alpha\beta} + c_\alpha^\gamma x^i_{,\gamma} x^i_{,\beta} = h_{\alpha\beta} + c_\alpha^\gamma g_{\gamma\beta} = 0,$$

i.e., $c_\alpha^\gamma = -h_\alpha^\gamma$, which means that

$$n^i_{,\gamma} = -h_\gamma^\beta x^i_{,\beta}. \tag{4.11}$$

Relation (4.11) is called the "Weingarten formula." It will be used in Chapter 5.

The derivatives $x^i_{,\alpha\beta}$ also play an important role. The vector $\mathbf{x}_{,\alpha\beta}$ can be decomposed into a component in the tangent plane and a component parallel to **n**. With some coefficients $\Gamma_{\alpha\beta}{}^{\gamma}$ we have

$$x^i_{,\alpha\beta} = \Gamma_{\alpha\beta}{}^{\gamma} x^i_{,\gamma} + h_{\alpha\beta} n^i. \tag{4.12}$$

The $\Gamma_{\alpha\beta}{}^{\gamma}$ are called "Christoffel symbols" of the second kind. The Christoffel symbols of the first kind are defined by

$$\Gamma_{\alpha\beta\gamma} := g_{\gamma\delta} \Gamma_{\alpha\beta}{}^{\delta}. \tag{4.13}$$

We can calculate the $\Gamma_{\alpha\beta\gamma}$ as follows: From

$$g_{\alpha\beta} = x^i_{,\alpha} x^i_{,\beta}$$

we derive

$$\begin{aligned}\frac{\partial g_{\alpha\beta}}{\partial y^{\gamma}} &= x^i_{,\alpha\gamma} x^i_{,\beta} + x^i_{,\alpha} x^i_{,\beta\gamma} \\ &= \Gamma_{\alpha\gamma}{}^{\delta} x^i_{,\delta} x^i_{,\beta} + \Gamma_{\beta\gamma}{}^{\delta} x^i_{,\alpha} x^i_{,\delta},\end{aligned}$$

so that

$$\Gamma_{\alpha\beta\gamma} = \Gamma_{\alpha\gamma}{}^{\delta} g_{\delta\beta} + \Gamma_{\beta\gamma}{}^{\delta} g_{\alpha\delta} = \Gamma_{\alpha\gamma\beta} + \Gamma_{\beta\gamma\alpha}. \tag{4.14}$$

Interchanging the indices, we find after some simple algebra that

$$\Gamma_{\alpha\beta\gamma} = \frac{1}{2}\left(\frac{\partial g_{\alpha\gamma}}{\partial y^{\beta}} + \frac{\partial g_{\beta\gamma}}{\partial y^{\alpha}} - \frac{\partial g_{\alpha\beta}}{\partial y^{\gamma}}\right). \tag{4.15}$$

It also follows from (4.12) and (4.13) that

$$\Gamma_{\alpha\beta\gamma} = x^i_{,\alpha\beta} x^i_{,\gamma}. \tag{4.16}$$

Since the metric tensor of S is symmetric, it follows that

$$\Gamma_{\alpha\beta\gamma} = \Gamma_{\beta\alpha\gamma} \qquad \text{and} \qquad \Gamma_{\alpha\beta}{}^{\gamma} = \Gamma_{\beta\alpha}{}^{\gamma}.$$

We can rewrite the expression for the curvature vector $\ddot{\mathbf{x}}$ in terms of the Christoffel symbols to get an expression for its tangential components.

$$\ddot{\mathbf{x}} \cdot \mathbf{x}_{,\gamma} = (x^i_{,\alpha\beta} \dot{y}^{\alpha} \dot{y}^{\beta} + x^i_{,\alpha} \ddot{y}^{\alpha}) x^i_{,\gamma} = \Gamma_{\alpha\beta\gamma} \dot{y}^{\alpha} \dot{y}^{\beta} + g_{\alpha\gamma} \ddot{y}^{\alpha}.$$

A curve C on S, whose curvature vector is normal to S at every point, is called a geodesic. By definition the geodesic curvature k_g of a geodesic is zero at every point. From our derivation above we get the differential equation of a geodesic (if s is the arc length along the geodesic) as

$$\Gamma_{\alpha\beta\gamma} \dot{y}^{\alpha} \dot{y}^{\beta} + g_{\alpha\gamma} \ddot{y}^{\alpha} = 0,$$

or in equivalent form as

$$\Gamma_{\alpha\beta}{}^{\gamma} \dot{y}^{\alpha} \dot{y}^{\beta} + \ddot{y}^{\gamma} = 0, \qquad \gamma = 1, 2.$$

This equation will recur later in a somewhat more general context.

4.3 RIEMANNIAN GEOMETRY AND TENSOR ANALYSIS

A necessary prerequisite for the understanding of this section is some familiarity with basic notions associated with tensors. In particular it is assumed that the reader knows what is meant, e.g., by "covariant," "contravariant," "contraction." Again, only the essentials that will be used in later chapters are introduced, which necessarily means a brief presentation of some selected topics.

4.3.1 Covariant Derivatives

Let g_{ik} be a symmetric positive definite covariant tensor in the variables $x^j, j = 1, \ldots, n$. A Riemannian space R^n is defined essentially by setting

$$ds^2 = g_{ik}\, dx^i\, dx^k \tag{4.17}$$

for the line element in R^n. The length L of a curve C given by

$$x^i = x^i(t), \qquad t_0 \leq t \leq t_1,$$

is therefore

$$L = \int_{t_0}^{t_1} \left(g_{ik} \frac{dx^i}{dt} \frac{dx}{dt} \right)^{1/2} dt.$$

The volume element dv in R^n is given by

$$dv = (\det g_{ik})^{1/2}\, dx^1\, dx^2 \cdots dx^n =: (g)^{1/2}\, dx^1\, dx^2 \cdots dx^n.$$

The scalar product $\mathbf{a} \cdot \mathbf{b}$ is defined by

$$\mathbf{a} \cdot \mathbf{b} = g_{ik} a^i b^k = a_k b^k = a^i b_i.$$

Here, the a^i are called the contravariant components of $\mathbf{a}$, and the a_i, the covariant components of $\mathbf{a}$. As in the previous section (in the context of surfaces) we define

$$g^{ik} = (g_{ik})^{-1},$$

and it is easy to check that g_{ik} is indeed a doubly contravariant tensor. It then follows that

$$a^i = g^{ik} a_k.$$

For an arbitrary tensor t^i_{jk}, e.g., we can "raise" or "lower" an index by multiplying either by g^{ik} or by g_{ik}. For example, the second covariant index k of the tensor t^i_{jk} is "raised" by setting $t^{im}_j = t^i_{jk} g^{km}$. The Christoffel symbols

Γ_{ijk} and $\Gamma_{ij}{}^{k}$ are now defined by

$$\Gamma_{ijk} := \frac{1}{2}\left(\frac{\partial g_{ik}}{\partial x^j} + \frac{\partial g_{ik}}{\partial x^i} - \frac{\partial g_{ij}}{\partial x^k}\right) \tag{4.18}$$

and

$$\Gamma_{ij}{}^{k} := g^{lk}\Gamma_{ijl}. \tag{4.19}$$

It is important to note that the Christoffel symbols are not tensors. Suppose we define a new set of local coordinates by

$$\tilde{x}^i = \tilde{x}^i(x^j), \qquad i, j = 1, \ldots, n. \tag{4.20}$$

Suppose that $\tilde{\Gamma}_{ij}{}^{k}$ are the Christoffel symbols in the new coordinates. Using the fact that g_{ik}, g^{ik} are tensors, an explicit calculation shows that

$$\Gamma_{ij}{}^{k} = \left(\tilde{\Gamma}_{pq}{}^{r}\frac{\partial \tilde{x}^p}{\partial x^i}\frac{\partial \tilde{x}^q}{\partial x^j} + \frac{\partial^2 \tilde{x}^r}{\partial x^i \partial x^j}\right)\frac{\partial x^k}{\partial \tilde{x}^r}. \tag{4.21}$$

Because of the term with second derivatives on the right-hand side of (4.21), it it obvious that the $\Gamma_{ij}{}^{k}$ are not tensors. Since $\Gamma_{ijk} = g_{kp}\Gamma_{ij}{}^{p}$ and g_{kp} is a tensor, it follows that Γ_{ijk} cannot be a tensor either. Likewise, one can check, e.g., that the partial derivative $\partial t^i/\partial x^k$ of a contravariant tensor of rank 1 (vector) is not a tensor. Our next aim is now to define a derivative of a tensor that will again be a tensor. More precisely, we want to define the covariant derivative of a tensor such that this new tensor is higher by one in its covariant rank. So, e.g., the covariant derivative $t^i_{,k}$ of t^i should be a mixed tensor of rank 2.

To find out how the covariant derivative $t^i_{,k}$ has to be defined, we differentiate the relation

$$\tilde{t}^i = \frac{\partial \tilde{x}^i}{\partial x^l} t^l.$$

This gives

$$\frac{\partial \tilde{t}^i}{\partial \tilde{x}^q}\frac{\partial \tilde{x}^q}{\partial x^p} = \frac{\partial \tilde{x}^i}{\partial x^l}\frac{\partial t^l}{\partial x^p} + \frac{\partial^2 \tilde{x}^i}{\partial x^l \partial x^p} t^l,$$

i.e.,

$$\frac{\partial \tilde{t}^i}{\partial \tilde{x}^q} = \frac{\partial \tilde{x}^i}{\partial x^l}\left(\frac{\partial \tilde{x}^q}{\partial x^p}\right)^{-1}\frac{\partial t^l}{\partial x^p} + \frac{\partial^2 x^i}{\partial x^l \partial x^p}\left(\frac{\partial \tilde{x}^q}{\partial x^p}\right)^{-1} t^l,$$

showing that the partial derivative of t^i is not a tensor because of the second derivative term on the right. But according to (4.21) we have

$$\frac{\partial^2 x^i}{\partial x^l \partial x^p} = \Gamma_{lp}{}^{s}\frac{\partial \tilde{x}^i}{\partial x^s} - \tilde{\Gamma}_{mn}{}^{i}\frac{\partial \tilde{x}^m}{\partial x^l}\frac{\partial \tilde{x}^n}{\partial x^p}$$

Inserting this relation into the expression for $(\partial \bar{t}^i/\partial \bar{x}^q)$ we find after some rearrangement that the quantity

$$t^i_{,k} := \frac{\partial t^i}{\partial x^k} + \Gamma_{jk}{}^i t^j \tag{4.22}$$

is indeed a mixed tensor of rank 2.

A similar calculation shows that for a covariant vector t_i the quantity

$$t_{i,k} := \frac{\partial t_i}{\partial x^k} - \Gamma_{ik}{}^s t_s \tag{4.23}$$

is a doubly covariant tensor. More generally, for an arbitrary tensor $t^{q_1,\ldots,q_s}_{l_1,\ldots,l_r}$ the covariant derivative has to be defined by

$$t^{q_1,\ldots,q_s}_{l_1,\ldots,l_r,i} = \frac{\partial t^{q_1,\ldots,q_s}_{l_1,\ldots,l_r}}{\partial x^i} + \sum_{n=1}^{s} t^{q_1,\ldots,q_{n-1}pq_{n+1},\ldots,q_s}_{l_1,\ldots,l_r} \Gamma_{pi}{}^{q_n}$$
$$- \sum_{m=1}^{r} t^{q_1,\ldots,q_s}_{l_1,\ldots,l_{m-1}\,kl_{m+1},\ldots,l_r} \Gamma_{il_m}{}^k. \tag{4.24}$$

For a scalar function $u(x^i)$, we set as before

$$u_{,i} = \frac{\partial u}{\partial x^i}, \tag{4.25}$$

which is easily seen to be a covariant vector.

Theorem 4.1 (Ricci) The covariant derivatives of g_{ik}, g^{ik}, δ^k_i are zero (in the sense of tensors).

Proof By definition we have

$$g_{ik,j} = \frac{g_{ik}}{\partial x^j} - g_{sk}\Gamma_{ij}{}^s - g_{is}\Gamma_{kj}{}^s$$
$$= \frac{\partial g_{ik}}{\partial x^j} - \Gamma_{ijk} - \Gamma_{kji}.$$

According to (4.18), however, the right-hand side in the preceding equation is zero. Next we also have by definition

$$\delta^i{}_{k,j} = \frac{\partial \delta^i_k}{\partial x^j} + \Gamma_{sj}{}^i \delta^s_k - \Gamma_{kj}{}^l \delta^i_l = \Gamma_{kj}{}^i - \Gamma_{kj}{}^i = 0.$$

Finally, we use the fact that

$$g_{ik}g^{kj} = \delta^j_i$$

and the fact, which is easily checked, that the ordinary rule for the differentiation of a product holds also for covariant differentiation. This yields

$$g_{ik,l}g^{kj} + g_{ik}g^{kj}{}_{,l} = \delta^j_{i,l} = 0,$$

and hence we also find that

$$g^{kj}{}_{,l} = 0.$$

It is useful to define also a contravariant derivative of say a tensor t^{ij}_k by

$$t^{ij,l}_k = g^{ls}t^{ij}_{k,s}, \tag{4.26}$$

i.e., we simply raise the index obtained from covariant differentiation. In this notation, we find, for instance,

$$u^{,i} = g^{ik}u_{,k} = g^{ik}\frac{\partial u}{\partial x^k},$$

for the contravariant components of the gradient of u.

4.3.2 Higher Derivatives, Curvatures

Let $u(x^i)$ be a scalar. By definition we have

$$u_{,i} = \frac{\partial u}{\partial x^i} \qquad \text{so that} \quad u_{,ij} = \frac{\partial^2 u}{\partial x^i \partial x^j} - \Gamma_{ij}{}^k u_{,k}.$$

Clearly the second covariant derivatives of a scalar are symmetric as is the case for ordinary derivatives. However, this is no longer true if we consider an arbitrary tensor. Consider, for instance, a covariant tensor t_k. Then

$$t_{k,i} = \frac{\partial t_k}{\partial x^i} - \Gamma_{ki}{}^p t_p,$$

$$\begin{aligned} t_{k,ij} &= \frac{\partial t_{k,i}}{\partial x^j} - \Gamma_{kj}{}^q t_{q,i} - \Gamma_{ki}{}^r t_{r,j} \\ &= \frac{\partial^2 t_k}{\partial x^i \partial x^j} - \frac{\partial t_s}{\partial x^j}\Gamma_{ki}{}^s - t_s\frac{\partial}{\partial x^j}\Gamma_{ki}{}^s - \left(\frac{\partial t_q}{\partial x^i} - t_s\Gamma_{qi}{}^s\right)\Gamma_{kj}{}^q \\ &\quad - \left(\frac{\partial t_r}{\partial x^j} - t_s\Gamma_{rj}{}^s\right)\Gamma_{ki}{}^r. \end{aligned}$$

We now form

$$\begin{aligned} t_{k,ij} - t_{k,ji} &= t_s\left\{\frac{\partial}{\partial x^i}\Gamma_{kj}{}^s - \frac{\partial}{\partial x^j}\Gamma_{ki}{}^s + \Gamma_{li}{}^s\Gamma_{kj}{}^l - \Gamma_{lj}{}^s\Gamma_{ki}{}^l\right\} \\ &= t_s R^s{}_{kij}. \end{aligned} \tag{4.27}$$

Since $t_{k,ij}$ and $t_{k,ji}$ are tensors, it follows that $R^s{}_{kij}$ must be a tensor as well. It is called the "mixed curvature tensor." The tensor

$$R_{lkij} := g_{sl} R^s{}_{kij} \tag{4.28}$$

is called the "curvature tensor" of the space R^n.

It is not hard to derive an analogous identity, called a *Ricci identity*, for other types of tensors. Since the covariant derivatives of the metric tensors are zero, we have, e.g.,

$$t^k{}_{,ij} = g^{kl} t_{l,ij};$$

thus

$$t^k{}_{,ij} - t^k{}_{,ji} = g^{kl}(t_{l,ij} - t_{l,ji}) = g^{kl} R^p{}_{lij} t_p = g^{kl} g^{pq} R_{qlij} t_p. \tag{4.29}$$

From our relations (4.27) and (4.28), it follows immediately that

$$R_{qlij} = R_{lqji} = -R_{lqij} = -R_{qlji} = R_{ijql}. \tag{4.30}$$

Interchanging the order of the indices q and l in the last term in (4.29), we now see by means of (4.30) that

$$t^k{}_{,ij} - t^k{}_{,ji} = -t^s R^k{}_{sij}. \tag{4.31}$$

For a general tensor $t^{k_1,\dots,k_s}_{m_1,\dots,m_r}$ the Ricci identity can be derived analogously, and then reads

$$t^{k_1,\dots,k_s}_{m,\dots,m_r,ij} - t^{k_1,\dots,k_s}_{m,\dots,m_r,ji} = \sum_{p=1}^{r} t^{k_1,\dots,k_s}_{m_1,\dots,m_{p-1}nm_{p+1},\dots,m_r} R^n{}_{m_p ij}$$
$$- \sum_{q=1}^{r} t^{k_1,\dots,k_{q-1}nk_{q+1},\dots,k_s}_{m_1,\dots,m_r} R^{k_q}{}_{nij}. \tag{4.32}$$

In view of our later applications we note that for the gradient of a scalar u the Ricci identity reduces to

$$u_{,kil} - u_{,kji} = R^s{}_{kij} u_{,s}. \tag{4.33}$$

From the curvature tensor new tensors and invariants can be formed. For instance, the tensor

$$R_{ki} := R^s{}_{kis} = g^{pi} R_{pkij} \tag{4.34}$$

is called the *Ricci tensor*, which plays an important role in a number of different contexts (see also Chapter 8). Note that $R_{ki} = R_{ik}$. The Ricci tensor can be calculated from the metric tensor and its derivatives. A calculation yields

$$R_{hj} = \frac{\partial^2 \log\sqrt{g}}{\partial x^h\, \partial x^j} - \frac{\partial}{\partial x^i} \Gamma_{hj}{}^i + \Gamma_{hi}{}^p \Gamma_{pj}{}^i - \Gamma_{hj}{}^p \frac{\partial \log\sqrt{g}}{\partial x^p}. \tag{4.35}$$

Let **e** be a vector of unit length. Then the invariant

$$K_R := R_{nj}e^n e^j \tag{4.36}$$

is called the *Ricci curvature* of the space R^n in the direction **e**. The scalar

$$R := g^{hj}R_{hj} \tag{4.37}$$

is called the "scalar curvature" of R^n. It is not hard to check that for n orthogonal vectors $\mathbf{e}_{(p)}$ of unit length we have

$$R = -\sum_{p=1}^{n} R_{hj}e^h_{(p)}e^j_{(p)}, \tag{4.38}$$

i.e., the sum on the right in (4.38) is independent of our choice of the directions $\mathbf{e}_{(p)}$.

In the special case in which $g_{ik} = g_{\alpha\beta}$, i.e., g_{ik} is the metric tensor of a surface S in E^3, the curvature tensor degenerates. Because of the relation (4.30), most of the components of the curvature tensor vanish. The only nonzero components turn out to be

$$R_{1212} = R_{2121}, \qquad R_{2112} = R_{1221}, \qquad R_{1212} = -R_{2112}.$$

Therefore it follows that for the Ricci tensor

$$R_{11} = g^{22}R_{2112}, \qquad R_{12} = R_{21} = 0, \qquad R_{22} = g^{11}R_{1221}.$$

We can relate the Ricci tensor to higher derivatives of the position vector of a point x^i of S as follows. By (4.33) we have

$$\begin{aligned}(x^i_{,\alpha\beta\gamma} - x^i_{,\alpha\gamma\beta})x^{i,\gamma} &= (x^i_{,\delta}R^\delta_{\alpha\beta\gamma})x^{i,\gamma} = x^i_{,\delta}x^{i,\gamma}R^\delta{}_{\alpha\beta\gamma} \\ &= x^i_{,\delta}x^i_{,\varepsilon}g^{\varepsilon\gamma}R^\delta{}_{\alpha\beta\gamma} = g^{\varepsilon\gamma}R_{\varepsilon\alpha\beta\gamma} = R_{\alpha\beta}.\end{aligned} \tag{4.39}$$

Note that in this equation *covariant* derivatives with respect to $g_{\alpha\beta}$ are used. In this sense we therefore have

$$x^i_{,\alpha\beta} = n^i h_{\alpha\beta}.$$

Inserting this last relation into (4.39), we find

$$g_{\delta\gamma}R^\delta{}_{\alpha\beta\gamma} = (h_{\alpha\beta}n^i_{,\gamma} - h_{\alpha\gamma}n^i_{,\beta})x^i_{,\gamma}.$$

At this point the Weingarten formula (4.11) can be applied, thus yielding

$$g_{\delta\gamma}R^\delta{}_{\alpha\beta\gamma} = (h_{\alpha\beta}h^\varepsilon_\gamma x^i_{,\varepsilon} - h_{\alpha\gamma}h^\nu_\beta x^i_{,\nu})x^i_{,\gamma} = h_{\alpha\beta}h^\varepsilon_\gamma g_{\varepsilon\gamma} - h_{\alpha\gamma}h^\nu_\beta g_{\nu\gamma},$$

so that

$$R_{\gamma\alpha\beta\gamma} = h_{\alpha\beta}h_{\gamma\gamma} - h_{\alpha\gamma}h_{\beta\gamma}.$$

In particular it follows that

$$\frac{R_{1212}}{g} = \frac{\det(h_{\alpha\beta})}{\det(g_{\alpha\beta})} = \det(h^{\beta}_{\alpha}) = K_G. \tag{4.40}$$

Similarly one finds that

$$R = 2K_G \tag{4.41}$$

in the case of a surface S in E^3.

4.3.3 Hypersurfaces

Similar to a surface S in Euclidean space E^3, we may consider hypersurfaces S^{n-1} of the Riemannian space R^n. Such a hypersurface may be given in parametric representation by

$$x^i = x^i(y^\alpha), \qquad i = 1, \ldots, n; \qquad \alpha = 1, \ldots, n-1. \tag{4.42}$$

In the following calculations Greek indices will always run from 1 to $n - 1$, italic indices from 1 to n, and the summation convention will be used for both kinds of indices.

The metric of R^n induces a metric in S^{n-1}. In R^n the line element is given by

$$ds^2 = g_{ik}\, dx^i\, dx^k.$$

But in S^{n-1} we have

$$dx^i = x^i_{,\alpha}\, dy^\alpha,$$

and therefore in S^{n-1}

$$ds^2 = g_{ik} x^i_{,\alpha} x^k_{,\beta}\, dy^\alpha\, dy^\beta := g_{\alpha\beta}\, dy^\alpha\, dy^\beta. \tag{4.43}$$

It follows that $g_{\alpha\beta}$ is a positive definite symmetric tensor, the induced metric tensor of S^{n-1}. Note that in the Euclidean case, i.e., when $g_{ik} = \delta_{ik}$, we are led back to the relation

$$g_{\alpha\beta} = x^i_{,\alpha} x^i_{,\beta}$$

as it was defined for the case $n = 3$ in Section 4.2.

The contravariant metric tensor is defined as before by

$$g_{\alpha\gamma} g^{\beta\gamma} = \delta^\beta_\alpha, \tag{4.44}$$

and the Christoffel symbols according to (4.15) by

$$\Gamma_{\alpha\beta\gamma} = \frac{1}{2}\left\{\frac{\partial g_{\alpha\gamma}}{\partial y^\beta} + \frac{\partial g_{\beta\gamma}}{\partial y^\alpha} - \frac{\partial g_{\alpha\beta}}{\partial y^\gamma}\right\}, \tag{4.45}$$

$$\Gamma_{\alpha\beta}{}^{\gamma} = g^{\gamma\delta} \Gamma_{\alpha\beta\delta}. \tag{4.46}$$

The Christoffel symbols for S^{n-1} can be calculated from the Christoffel symbols of R^n. A direct computation gives

$$\Gamma_{\alpha\beta\gamma} = \left(\Gamma_{ijk}x^i_{,\alpha}x^j_{,\beta} + g_{jk}\frac{\partial^2 x^j}{\partial y^\alpha\,\partial y^\beta}\right)x^k_{,\gamma}. \tag{4.47}$$

We can now repeat some of our earlier calculations in order to extend the formulas of Section 4.3.2, valid for a surface in E^3 to hypersurfaces S^{n-1} in R^n.

Let n^i be a unit normal vector to S^{n-1}. By assumption we have

$$g_{ik}n^i n^k = 1, \tag{4.48}$$

and also

$$g_{ik}n^i x^k_{,\alpha} = 0, \qquad \alpha = 1, \ldots, n-1. \tag{4.49}$$

We differentiate (4.48) covariantly with respect to y^β, remembering that g_{ik} and n^i have to be treated like scalars in this case. This gives

$$\frac{\partial g_{ik}}{\partial x^l}x^l_{,\beta}n^i n^k + 2g_{ik}n^i_{,\beta}n^k = 0$$

where

$$x^l_{,\beta} = \frac{\partial x^l}{\partial y^\beta}, \qquad n^k_{,\beta} = \frac{\partial n^k}{\partial y^\beta}.$$

By definition

$$\frac{\partial g_{ik}}{\partial x^l} = \Gamma_{ilk} + \Gamma_{kli},$$

so that

$$\frac{\partial g_{ik}}{\partial x^l}n^i n^k = (\Gamma_{ilk} + \Gamma_{kli})n^i n^k.$$

By interchanging indices we arrive at

$$n^j g_{jk}(n^j_{,\beta} + \Gamma_{lp}{}^k n^p x^l_{,\beta}) = 0,$$

which means that the vector in the braces is tangential, i.e.,

$$n^k_{,\beta} + \Gamma_{lp}{}^k n^p x^l_{,\beta} = C^\gamma_\beta x^k_{,\gamma},$$

for some scalar C^γ_β. To determine C^γ_β we also take covariant derivatives in (4.49), which leads to

$$\frac{\partial g_{ik}}{\partial x^l}x^l_{,\beta}n^i x^k_{,\alpha} + g_{ik}(n^i_{,\beta}x^k_{,\alpha} + n^i x^k_{,\alpha\beta}) = 0.$$

The derivatives of g_{ik} can again be replaced by Christoffel symbols. After some rearrangement we find

$$C^{\gamma}_{\beta} = -h^{\gamma}_{\beta} = -g^{\gamma\alpha}h_{\alpha\beta}, \qquad \text{where} \quad h_{\alpha\beta} := x^{i}_{,\alpha\beta}n^{i}$$

is the second fundamental tensor of S^{n-1}. The "Weingarten formulas" take now the form

$$n^{k}_{,\beta} = -h^{\gamma}_{\beta}x^{k}_{,\gamma} - \Gamma_{lp}{}^{k}n^{p}x^{l}_{,\beta}. \tag{4.50}$$

4.3.4 Intrinsic Derivatives and Generalized Covariant Differentiation

Let C be a curve in R^n, given in the parametric form

$$x^i = x^i(s),$$

where s measures the arc length (in R^n!) along C. Let t^{ik} be some contravariant tensor. Then along C

$$\frac{dt^{ik}}{ds} = \frac{\partial t^{ik}}{\partial x^j}\dot{x}^j.$$

It is clear that $\dot{x}^j$ is a contravariant vector. However, as we have already seen, $\partial t^{ik}/\partial x^j$ is not a tensor, and therefore neither is dt^{ik}/ds. But if we define the intrinsic derivative D/ds by

$$\frac{Dt^{ik}}{ds} := t^{ik}_{,j}\dot{x}^j = \frac{dt^{ik}}{ds} + (\Gamma_{pj}{}^{i}t^{pk} + \Gamma_{qj}{}^{k}t^{iq})\dot{x}^j, \tag{4.51}$$

then (Dt^{ik}/ds) is obviously a tensor. It is clear how the intrinsic derivative is defined for an arbitrary tensor. For a scalar u the rule (4.51) simply gives

$$\frac{Du}{ds} = \frac{du}{ds}. \tag{4.52}$$

Another interesting case is the following. For the contravariant vector $\dot{x}^i$ we find

$$\frac{D\dot{x}^i}{ds} = \ddot{x}^i + \Gamma_{jk}{}^{i}\dot{x}^j\dot{x}^k. \tag{4.53}$$

A curve C is called "autoparallel" if $D\dot{x}^i/ds = 0$ (i.e., if the tangent vectors are parallel in the sense of intrinsic differentiation). According to (4.53), C is autoparallel if and only if

$$\ddot{x}^i + \Gamma_{jk}{}^{i}\dot{x}^j\dot{x}^k = 0, \qquad i = 1, \ldots, n. \tag{4.54}$$

On a surface S of E^3 Eq. (4.54) corresponds to the differential equation of a geodesic if the italic indices are exchanged for Greek ones.

There is an analog to the intrinsic differentiation in the theory of hypersurfaces of R^n. Consider, e.g., a tangent vector $x^i_{,\alpha}$ to a surface S^{n-1}. Then the derivative $\partial x^i_{,\alpha}/\partial y^\beta$ is not a tensor with respect to transformations of the y^α system. We wish to define a differentiation such that for a given tensor of a mixed character, say t^{ik}_{α}, we can take a "generalized covariant derivative" $t^{ik}_{l;\beta}$, and such that the derived quantity is a tensor of the kind indicated by the position of the Greek and italic indices. It is easy to check that with the following definition we can satisfy these requirements:

$$\begin{aligned} t^{ik}_{l\alpha;\beta} &= t^{ik}_{l\alpha,j}x^j_{,\beta} - \Gamma_{\alpha\beta}{}^{\gamma}t^{ik}_{l\gamma} \\ &= \frac{\partial t^{ik}_{l\alpha}}{\partial y^\beta} + (\Gamma_{pj}{}^{i}t^{pk}_{l\alpha} + \Gamma_{pj}{}^{k}t^{ip}_{l\alpha} - \Gamma_{lj}{}^{r}t^{ik}_{r\alpha})x^j_{,\beta} - \Gamma_{\alpha\beta}{}^{\gamma}t^{ik}_{l\gamma}. \end{aligned} \tag{4.55}$$

It is again clear how the rule (4.55) has to be extended for an arbitrary tensor. In the special case of a tensor with only italic indices, say t^{ik}, (4.55) reduces to

$$t^{ik}_{l;\beta} = t^{ik}_{l,j}x^j_{,\beta} = \frac{\partial t^{ik}_{l}}{\partial y^\beta} + (\Gamma_{pj}{}^{i}t^{pk}_{l} + \Gamma_{pj}{}^{k}t^{ip}_{l} - \Gamma_{lj}{}^{p}t^{ik}_{p})x^j_{,\beta}. \tag{4.56}$$

On the other hand (4.55) reduces to

$$t_{\alpha;\beta} = \frac{\partial t_\alpha}{\partial y^\beta} - \Gamma_{\alpha\beta}{}^{\gamma}t_\gamma = t_{\alpha,\beta} \tag{4.57}$$

for a tensor with only Greek indices, say t_α. Equation (4.57) is the old rule for covariant differentiation with respect to the metric S^{n-1}.

In the previous section the generalized Weingarten formula stated that

$$n^k_{,\beta} = -h^\gamma_\beta x^{\ k}_{,\gamma} - \Gamma_{lp}{}^{k}n^p x^l_{,\beta}.$$

We can rewrite this formula by applying the rule for generalized covariant differentiation. Then it simplifies to

$$n^k_{;\beta} = -h^\gamma_\beta x^k_{,\gamma}. \tag{4.58}$$

Similarly, the formulas of Gauss can now be written as

$$x^i_{;\alpha\beta} = h_{\alpha\beta}n^i. \tag{4.59}$$

4.3.5 Laplacian and Gradient

The gradient in R^n of a scalar u is to be defined such that $|\nabla u|^2$ is a scalar and therefore an invariant. In addition, the definition ought to give the original gradient in the Euclidean case. Thus we define

$$|\nabla u|^2 := g^{ik}u_{,i}u_{,k} = u^{,k}u_{,k}$$

and call $u_{,k}$ and $u^{,k}$ the covariant and contravariant components of the gradient, respectively. Given an arbitrary unit vector $\mathbf{e}$ we define the directional derivative $\partial u/\partial e$ of a scalar u as

$$\frac{\partial u}{\partial e} := (\nabla u, \mathbf{e}) = g^{ik}u_{,i}e_k = u_{,i}e^i = u^{,k}e_k. \tag{4.60}$$

The following inequality is a straightforward extension of the Euclidean analog, namely,

$$\left(\frac{\partial u}{\partial e}\right)^2 \leq |\nabla u|^2. \tag{4.61}$$

For the proof of (4.62) we start with the obvious inequality

$$g^{ik}(u_{,i} + \varepsilon e_i)(u_{,k} + \varepsilon e_k) \geq 0,$$

where ε is an arbitrary scalar (number). The inequality can be written as

$$g^{ik}u_{,i}u_{,k} + 2\varepsilon g^{ik}e_i u_{,k} + \varepsilon^2 g^{ik}e_i e_k \geq 0,$$

i.e.,

$$|\nabla u|^2 + 2\varepsilon e^k u_{,k} + \varepsilon^2 \geq 0.$$

Using the fact that the discriminant of the quadratic expression in ε must be negative, we are immediately led to (4.61).

As a next step the Laplacian in R^n is to be defined such that it coincides with the ordinary Laplacian in the case in which $g_{ik} = \delta_{ik}$. Furthermore, the quantity Δu, where u is a scalar, has to be an invariant in R^n. This leads to the definition

$$\Delta u := g^{ik}u_{,ik}, \tag{4.62}$$

where *covariant* derivatives are used.

If $g_{ik} = g_{\alpha\beta}$ is the metric tensor of a surface, Eq. (4.62) is to be interpreted accordingly. In the special case of a surface S^{n-1} in Euclidean space E^n, (4.62) can be written in terms of ordinary derivatives. An explicit calculation gives the Laplacian in the more familiar form

$$\Delta u = \frac{1}{\sqrt{g}}\frac{\partial}{\partial y^\alpha}\left(\sqrt{g}\, g^{\alpha\beta}\frac{\partial u}{\partial y^\beta}\right). \tag{4.63}$$

For later applications, it is useful to remember the following rules whose proofs are straightforward

$$\Delta(uv) = u\,\Delta v + v\,\Delta u + 2\nabla u \cdot \nabla v, \tag{4.64}$$

where of course $\nabla u \cdot \nabla v := g^{ik}u_{,i}v_{,k} = u^{,k}v_{,k} = u_{,i}v^{,i}$. Furthermore, it follows that

$$\Delta(f(u)) = \frac{df}{du}\,\Delta u + \frac{d^2 f}{du^2}\,|\nabla u|^2, \tag{4.65}$$

as in the Euclidean case.

Let S^{N-1} be a hypersurface in R^N and n^i a unit normal to S^{N-1}. We define the first and second normal derivatives by

$$\frac{\partial u}{\partial n} := \nabla u \cdot \mathbf{n} = g^{ik}u_{,i}n_k = u_{,i}n^i, \tag{4.66}$$

and in accordance with Section 4.1

$$\frac{\partial^2 u}{\partial n^2} := u_{,ik}n^i n^k. \tag{4.67}$$

Let $\Delta_S u := g^{\alpha\beta}u_{,\alpha\beta}$ be the Laplacian in the induced metric of S^{N-1} and Δu the Laplacian of R^N. At any point of S^{N-1} the following relation holds:

$$\Delta u = \Delta_s u + (N-1)H\,\frac{\partial u}{\partial n} + \frac{\partial^2 u}{\partial n^2}. \tag{4.68}$$

To prove relation (4.68) we write on S^{N-1}

$$u = u(x^i(y^\alpha))$$

and calculate

$$u_{;\alpha\beta} = u_{,\alpha\beta} = u_{,ij}x^i_{,\alpha}x^j_{,\beta} + u_{,i}x^i_{;\alpha\beta},$$

and

$$\Delta_s u = g^{\alpha\beta}u_{,\alpha\beta} = u_{,ij}g^{\alpha\beta}x^i_{,\alpha}x^j_{,\beta} + u_{,i}g^{\alpha\beta}x^i_{;\alpha\beta}.$$

At this point, the formulas of Gauss as given in Eq. (4.59) can be applied. The normal n on S^{N-1} and the mean curvature H are now defined such that for a sphere of radius R we have

$$\frac{\partial}{\partial n} = \frac{\partial}{\partial r}, \qquad H = \frac{1}{R},$$

r measuring the distance from the center of the sphere. With this sign convention and (4.59), our last relation can be cast into the form

$$\Delta_s u = u_{,ij}g^{\alpha\beta}x^i_{,\alpha}x^j_{,\beta} - \frac{\partial u}{\partial n}(N-1)H.$$

Thus we find

$$\Delta_s u + (N-1)H\frac{\partial u}{\partial n} + \frac{\partial^2 u}{\partial n^2} = u_{,ij}(g^{\alpha\beta}x^i_{,\alpha}x^j_{,\beta} + n^i n^j).$$

It remains to show that on S^{N-1} one has

$$g^{\alpha\beta}x^i_{,\alpha}x^j_{,\beta} + n^i n^j = g^{ij}. \tag{4.69}$$

To verify Eq. (4.69) note that for any function $v(x^i)$ we have

$$\begin{aligned} g^{\alpha\beta}x^i_{,\alpha}x^j_{,\beta}v_{,i}v_{,j} + v_{,i}v_{,j}n^i n^j &= g^{\alpha\beta}v_{,\alpha}v_{,\beta} + v_{,i}v_{,j}n^i n^j \\ &= |\nabla_s v|^2 + \left(\frac{\partial v}{\partial n}\right)^2 = g^{ij}v_{,i}v_{,j}. \end{aligned} \tag{4.70}$$

Here ∇_s denotes the gradient in S^{N-1}. Clearly, (4.70) implies the validity of (4.69), and this completes the proof of (4.68).

Chapter 5

THE *P*-FUNCTION FOR SOLUTIONS OF $\Delta u + f(u) = 0$

The introduction of the function

$$P := g(u)|\nabla u|^2 + h(u),$$

where u is a solution of $\Delta u + f(u) = 0$, is an essential step for the development of all the following chapters. The main feature of the function P is that it satisfies a maximum principle if the functions $g(u)$, $h(u)$ are chosen appropriately. These maximum principles are then the starting point for the derivation of numerous useful bounds for all kinds of quantities that are of interest in this problem.

It is an important fact that this procedure can be extended in various directions for generalizations of the equation $\Delta u + f(u) = 0$. It seems convenient to examine first the simplest cases and some of their applications, and then proceed to more complicated situations. We therefore begin with the one-dimensional case in the first section.

5.1 THE ONE-DIMENSIONAL PROBLEM

If u is a function of a single variable x, our equation can be written as

$$u'' + f(u) = 0, \tag{5.1}$$

where the primes obviously denote differentiation with respect to x, and x lies in some interval (a, b). A thorough treatment of (5.1) can be found in many textbooks and original articles (see, e.g., Laetsch (1970)). The usual steps are as follows: Set

$$F(u) = \int_0^u f(s)\,ds.$$

Then a multiplication of (5.1) by u' gives

$$u''u' + f(u)u' = (\tfrac{1}{2}u'^2)' + (F(u))' = 0,$$

i.e.,

$$\tfrac{1}{2}u'^2 + F(u) = \text{const.} \tag{5.2}$$

From (5.2) one can easily derive an implicit representation of the solution, which we shall discuss briefly afterward. First, however, note that in the case of Eq. (5.1) we have thus found that the function

$$P := u'^2 + 2F(u)$$

is just a constant.

Let us now assume that the interval under consideration is finite and symmetric, say $(-d, d)$. Suppose that u vanishes at the end points. Also, it is useful to introduce a numerical parameter λ and take $\lambda f(u)$ in the place of $f(u)$. From the corresponding equation (5.2), we then deduce the following implicit representation of the solution:

$$\int_0^{u(x)} (F(u_m) - F(s))^{-1/2}\,ds = x(2\lambda)^{1/2}, \qquad 0 < x < d, \tag{5.3}$$

with $u(0) = u_m$, $u(-x) = u(x)$. For given values of d and λ we then must have

$$\Phi(u_m) := \int_0^{u_m} (F(u_m) - F(s))^{-1/2}\,ds = d(2\lambda)^{1/2}. \tag{5.4}$$

It is not hard to see that in turn we can define $u(x)$ by (5.3) (under mild qualifications for f) and get a solution of (5.1). It can be shown (see Laetsch, (1970)) that if $f(0) > 0$, $f' > 0$ and $f'' > 0$ for positive argument, then (5.4) can have a positive solution (and therefore (5.1) with f replaced by λf) only for $\lambda \in (0, \lambda^*)$, where λ^* is some finite positive value.

Similarly, as above different boundary conditions for $u(x)$ can be treated. Only condition (5.4) has to be replaced then. Note that in the special case in which $f(u) = \lambda u$, we can write (5.4) as

$$\Phi(u_m) := \int_0^{u_m} \frac{ds}{\sqrt{u_m^2 - s^2}} = d\sqrt{\lambda}.$$

But $\Phi(u_m) \equiv \frac{1}{2}\pi$ in this case, and hence $\lambda = (\pi/2d)^2$, corresponding to the first eigenvalue of $u'' + \lambda u = 0$ with zero-boundary data. In general, the function $\Phi(u_m)$ takes a complicated form: for instance, for $f(u) = e^u$ one would get

$$\Phi(u_m) = e^{-u_m/2}(u_m + 2\log(1 + (1 - e^{-u_m})^{1/2}).$$

Equations that are more general than (5.1) can be treated the same way. Take, e.g., the equation

$$h(u'^2)u'' + g(u) = 0. \tag{5.5}$$

Introducing $p(u) := u'^2$ and using the fact that

$$\frac{dp}{dx} = \frac{dp}{du}\frac{du}{dx} = 2\frac{du}{dx}\frac{d^2u}{dx^2}, \tag{5.6}$$

we find as before that

$$\tfrac{1}{2}H(u'^2) + G(u) = \text{const}, \tag{5.7}$$

where $dH/ds = h(s)$, $dG/ds = g(s)$. From (5.7) one can again derive an implicit representation of the solution of (5.5). The main reason for considering problems (5.1) and (5.5) is that they give an idea of what types of P-functions one has to look for in the n-dimensional versions of (5.1) or (5.5) or even more general forms such as, e.g.,

$$(h(u'^2, u)u')' + g(u, u'^2) = 0 \tag{5.8}$$

and their n-dimensional versions.

5.2 DETERMINATION OF P-FUNCTIONS FOR SOLUTIONS OF $\Delta u + f(u) = 0$

In all the following sections we shall assume that u is at least a $C^2(\bar{D}) \cap C^3(D)$ solution of

$$\Delta u + f(u) = 0 \qquad \text{in} \quad D, \tag{5.9}$$

where D is a finite domain in n-dimensional Euclidean space. Also, we shall restrict our attention to positive solutions. In many calculations the case $N = 2$ allows a somewhat different treatment than that of $N \geq 3$. Also, the difficulties that one will encounter depend for a good part on the boundary conditions imposed on u. We therefore split Section 5.2 into two subsections.

Before we consider the first case, the goal has to be clearly stated: We wish to find conditions under which the function $P = g(u)|\nabla u|^2 + h(u)$, u being a solution of (5.9), satisfies a maximum principle. According to Section 5.1, the function $P = |\nabla u|^2 + 2F(u)$ is a possible candidate.

We denote partial derivatives by a comma followed by a subscript, i.e., the same way that covariant derivatives were denoted in Chapter 4. Then

$$P_{,k} = g'(u)|\nabla u|^2 u_{,k} + 2gu_{,ik}u_{,i} + h'u_{,k}, \tag{5.10}$$

which shows that P may assume its maximum at a point at which $\nabla u = 0$. The result will be proved later. A second possibility is that P assumes its maximum somewhere on the boundary ∂D of D. The third possibility is that P assumes its maximum at an interior point of D, not a critical point of u, but a point where the determinant of the matrix

$$C_{ik} := 2gu_{,ik} + (h' + g'|\nabla u|^2)\,\delta_{ik}$$

vanishes. This third possibility does not lead to any useful information for our purposes, so we shall try to choose P in such a way that one of the first two cases mentioned occur.

5.2.1 The Two-Dimensional Case

The calculations of Section 5.1 suggest the choice of $P = |\nabla u|^2 + 2F(u)$, the higher dimensional analog of the expression given in Eq. (5.2). We go one step further and take the function

$$P = |\nabla u|^2 g(u) + h(u),$$

u being a solution of (5.9). Then

$$P_{,i} = 2u_{,ji}u_{,j}g + |\nabla u|^2 g' u_{,i} + h' u_{,i}, \tag{5.11}$$

and

$$\begin{aligned}\Delta P = P_{,ii} = {} & 2u_{,ij}u_{,ij}g + 2u_{,jii}u_{,j}g + 4u_{,j}u_{,i}u_{,ij}g' \\ & + |\nabla u|^4 g'' + |\nabla u|^2(g'u_{,ii} + h'') + h'u_{,ii}.\end{aligned} \tag{5.12}$$

Here a prime is used for derivatives with respect to u. Now

$$u_{,jii} = u_{,iij} = -f'u_{,j},$$

and the third term on the right in (5.12) can be expressed as

$$4u_{,j}u_{,i}u_{,ij}g' = 2(P_{,i} - |\nabla u|^2 g' u_{,i} - h' u_{,i})\frac{g'}{g}\, u_{,i}. \tag{5.13}$$

This allows us to rewrite (5.12) as

$$\begin{aligned}\Delta P = {} & |\nabla u|^4\left(g'' - 2\frac{(g')^2}{g}\right) + |\nabla u|^2\left(h'' - fg' - 2f'g - 2\frac{h'g'}{g}\right) \\ & + 2u_{,ij}u_{,ij}g - h'f + 2\frac{g'}{g}\, u_{,i}P_{,i}.\end{aligned} \tag{5.14}$$

Up to this point the calculation is the same in any number of dimensions. But in order to eliminate the term $2u_{,ij}u_{,ij}g$, we use an identity that is only valid in two dimensions. For any sufficiently smooth function v, we have

$$|\nabla v|^2 v_{,ij} v_{,ij} = |\nabla v|^2 (\Delta v)^2 + 2 v_{,i} v_{,ik} v_{,j} v_{,jk} - 2 \Delta v\, v_{,i} v_{,j} v_{,ij}. \tag{5.15}$$

The identity is easily checked by an explicit calculation. Equation (5.12) and the identity (5.15) yield

$$2g u_{,ij} u_{,ij} = 2 \frac{g}{|\nabla u|^2} \left\{ |\nabla u|^2 + \frac{1}{2g^2} (P_{,i} - g'|\nabla u|^2 u_{,i} - h' u_{,i}) \right.$$
$$\left. \times (P_{,i} - g'|\nabla u|^2 u_{,i} - h' u_{,i}) u_{,i} \right\}. \tag{5.16}$$

The combination of (5.14) and (5.16) finally leads to

$$\Delta P + \frac{L_i P_{,i}}{|\nabla u|^2} = g(\log g)''|\nabla u|^4 + ((h' - 2fg)' - fg')|\nabla u|^2$$
$$+ \frac{1}{g}(h' - fg)(h' - 2fg), \tag{5.17}$$

where

$$L_i = -\frac{1}{g}(P_{,i} - 2u_{,i}(h' - fg)).$$

In view of our aim, namely, to choose g, h such that P satisfies a maximum principle, it is evident that we have to select the functions g, h such that the quadratic form in $|\nabla u|^2$ in (5.17) is positive semidefinite. An obvious choice is, e.g.,

$$g = 1, \qquad h' = 2f,$$

i.e.,

$$P = |\nabla u|^2 + 2F(u). \tag{5.18}$$

as one may have conjectured already on the basis of Section 5.1. Two more examples are

$$g = e^{-cu}, \qquad h = 2 \int_0^u e^{-cs} f(s)\, ds, \qquad c > 0, \tag{5.19}$$

so that

$$P = e^{-cu}|\nabla u|^2 + 2 \int_0^u e^{-cs} f(s)\, ds.$$

If $f' \leq 0$, one may choose $g = 1$, $h' = f$, i.e.,

$$P = |\nabla u|^2 + F(u). \tag{5.20}$$

We have seen on the basis of (5.10) that P may assume its maximum at essentially three different places if g and h are chosen arbitrarily. However, if we restrict our choice to functions that make the quadratic form in (5.17) positive semidefinite, we are only left with the possibilities (for nonconstant P!)

(a) P assumes its maximum at a point where $\nabla u = 0$,
(b) P assumes its maximum on ∂D.

Let us state this result as

Lemma 5.1 Let u be a $C^3(D)$ solution of (5.9) and D a plane domain. If $g(u)$, $h(u)$ are chosen such that the quadratic form in (5.17) is positive semi-definite, then the corresponding function $P = g(u)|\nabla u|^2 + h(u)$ must assume its maximum on ∂D or at a critical point of u.

Proof By assumption, P satisfies

$$\Delta P + \frac{L_i P_{,i}}{|\nabla u|^2} \geq 0 \qquad \text{in} \quad D.$$

The maximum principle (see Chapter 3) can be applied at any point where $\nabla u \neq 0$, telling us that P must assume its maximum on ∂D, unless $P \equiv \text{const}$ (e.g., in the one-dimensional case with $g = 1$, $h = 2F$). The only other possibility is that P attains its maximum at at a point where $\nabla u = 0$, which proves Lemma 5.1.

Before investigating the two-dimensional case any further, let us see how one can derive an analog of (5.17) in more than two dimensions.

5.2.2 The Higher-Dimensional Case

The main problem still consists in eliminating the term $u_{,ij}u_{,ij}$ in (5.14). Since the identity (5.15) is only valid in two dimensions, we use Schwarz's inequality for vectors. From (5.11) it follows that

$$\begin{aligned}(P_{,i} - |\nabla u|^2 g' u_{,i} - h' u_{,i})&(P_{,i} - |\nabla u|^2 g' u_{,i} - h' u_{,i}) \\ &= 4u_{,ji}u_{,j}u_{,ki}u_{,k}g^2 \leq 4u_{,ij}u_{,ij}|\nabla u|^2 g^2. \end{aligned} \tag{5.21}$$

We can now combine inequality (5.21) with Eq. (5.14) to arrive at

$$\begin{aligned}\Delta P + \frac{L_k^* P_{,k}}{|\nabla u|^2} \geq &-2g^{3/2}(g^{-1/2})''|\nabla u|^4 + ((h' - 2fg)' + \frac{g'}{g}(fg - h'))|\nabla u|^2 \\ &+ \frac{h'}{2g}(h' - 2fg),\end{aligned} \tag{5.22}$$

where

$$L_k^* = \frac{1}{g}\left(h'u_{,k} - \frac{1}{2}P_{,k}\right) - |\nabla u|^2\frac{g'}{g}u_{,k}.$$

The analog of Lemma 5.1 now is

Lemma 5.2 Let u be a $C^3(D)$ solution of (5.9), $D \subset E^N$, $N \geq 3$. If $g(u)$, $h(u)$ are chosen such that the quadratic form in (5.22) is positive semidefinite, then the corresponding function $P = g(u)|\nabla u|^2 + h(u)$ must assume its maximum on ∂D or at a critical point of u.

The proof is of course the same as for Lemma 5.1.

So far we have proven that P satisfies an elliptic inequality everywhere in D. As noted by Payne (1976), it suffices to prove that P satisfies such an inequality at any point where $\nabla u = 0$. Payne in fact proved

Lemma 5.3 Let u be a $C^3(D)$-solution of (5.9) and $D \subset E^N$, $N > 1$. Let $g(u)$ satisfy

(i) $g(u) \geq 0$,
(ii) $f(u)g'(u) < 0 \quad$ for $\quad u \neq 0$,
(iii) $g(u)g''(u) - \dfrac{3N-4}{2(N-1)}(g'(u))^2 \geq 0$.

Then the function

$$P = g(u)|\nabla u|^2 + 2\int_0^u f(s)g(s)\,ds$$

assumes its maximum on ∂D or at a point where $\nabla u = 0$.

Proof We follow the proof given by Payne (1976). With our choice of g, h we have

$$P_{,k} = 2gu_{,i}u_{,ik} + g'|\nabla u|^2 u_{,k} + 2gfu_{,k}, \tag{5.23}$$

$$\begin{aligned}\Delta P &= 2gu_{,ik}u_{,ik} + 4g'u_{,i}u_{,k}u_{,ik} + g''|\nabla u|^4 + 2g'f|\nabla u|^2 - 2gf^2\\ &= 2gu_{,ik}u_{,ik} + 2(\log g)'u_{,k}P_{,k} + 2g''g'(\log g)'|\nabla u|^4 - 3g'f|\nabla u|^2 - 2gf^2.\end{aligned} \tag{5.24}$$

Suppose now that P takes its maximum at an interior point x at which $\nabla u \neq 0$. At x we can orient our axes such that $u_{,j}(x) = 0, j \neq 1$ and $u_{,1}(x) \neq 0$. Since ∇P must vanish at x, it follows that

$$2gu_{,11} + g'u_{,1}^2 + 2gf = 0 \tag{5.25a}$$

$$u_{,1i} = 0,\ i = 2, \ldots, N. \tag{5.25b}$$

From (5.25) we conclude that

$$u_{,11}^2 = \left(\frac{g'}{2g} u_{,1}^2 + f\right)^2, \tag{5.26}$$

and from the differential equation for u that

$$\Delta u - u_{,11} = \frac{g'}{2g} u_{,1}^2. \tag{5.27}$$

Schwarz's inequality gives

$$\sum_{i=2}^{N} u_{,ii}^2 \geq \frac{1}{N-1} (\Delta u - u_{,11})^2 = \frac{1}{N-1} \left(\frac{g'}{2g} u_{,1}^2\right)^2. \tag{5.28}$$

But

$$u_{,ij} u_{,ij} \geq u_{,kk} u_{,kk},$$

and hence

$$2g u_{,ij} u_{,ij} \geq \frac{N}{N-1} \frac{g'^2}{2g} u_{,1}^2 + 2g' u_{,1}^2 f + 2g f^2, \tag{5.29}$$

so that with (5.24), we find at x

$$\Delta P \geq \left(g'' - \frac{3N-4}{2(N-1)} \frac{g'^2}{g}\right) u_{,1}^2 - g' f u_{,1}^2. \tag{5.30}$$

Assumptions (i)–(iii) then imply

$$\Delta P > 0 \quad \text{at} \quad x, \tag{5.31}$$

contradicting the assumption that P takes a maximum there. This proves Lemma 5.3.

Remarks Lemma 5.2 implies that if $D \subset E^N$, $N > 2$, and $0 \leq \alpha \leq 2$, then the function

$$P = \frac{|\nabla u|^2}{(u+\beta)^\alpha} + 2 \int_0^u \frac{f(s)}{(s+\beta)^\alpha} ds, \qquad \beta > 0,$$

takes its maximum on ∂D or at a critical point of u. By Lemma 5.3 the same is still true if $0 \leq \alpha \leq 2(N-1)/(N-2)$.

So far we are still left with two possibilities as to where P may take its maximum. In the next section, we shall discuss the situation in which the maximum occurs on the boundary.

5.3 THE MAXIMUM OF P ON ∂D

5.3.1 Plane Domains

As before, the case of $N = 2$ dimensions allows a somewhat different treatment. It is useful to prove first a simple result which can be applied to more than two dimensions as well.

Lemma 5.4 For any sufficiently smooth function v the inequality

$$Nv_{,ik}v_{,ik} \geq (\Delta v)^2 \tag{5.32}$$

holds in N dimensions.

Proof Consider the quadratic form in ε

$$Q(\varepsilon) := (v_{,ik} - \varepsilon \Delta v \, \delta_{ik})(v_{,ik} - \varepsilon \Delta v \, \delta_{ik}) \geq 0.$$

The discriminant of $Q(\varepsilon)$ must be nonpositive, i.e.,

$$(\Delta v \, v_{,ik} \, \delta_{ik})^2 \leq v_{,ik} \, v_{,ik} (\Delta v)^2 \, \delta_{jl} \, \delta_{jl};$$

but

$v_{,ik} \, \delta_{ik} = \Delta v$ and $\delta_{jl} \, \delta_{jl} = N$, which gives inequality (5.32).

We can now prove

Theorem 5.1 Let u be a solution of (5.9) with $u_m \leq u \leq u_M$ and suppose that for $u_m \leq s \leq u_M$ the following conditions are satisfied:

(i) $(\log g(s))'' \geq 0, \qquad g(s) > 0,$
(ii) $(\log f(s))' + (2 \log g(s))' \leq 0.$

Then the function

$$P = g(u)|\nabla u|^2 + \int_0^u g(s) f(s) \, ds$$

assumes its maximum on ∂D.

Proof We have chosen $h(u)$ such that $h' = fg$. The quadratic form in $|\nabla u|^2$ in (5.17) is positive semidefinite with our assumptions (i) and (ii) if $h' = fg$. Note that assumption (ii) ensures the nonnegativity of the coefficient of $|\nabla u|^2$ in (5.17).

With our present choice of $g(u)$ and $h(u)$ the corresponding function P satisfies

$$\Delta P - \frac{1}{g} \frac{|\nabla P|^2}{|\nabla u|^2} \geq 0 \qquad \text{in} \quad D,$$

i.e.,

$$\Delta P \geq 0 \quad \text{in} \quad D.$$

The maximum principle for elliptic problems then implies the statement of Theorem 5.1.

Remarks Two examples of functions $g(u)$ satisfying the assumptions of Theorem 5.1 are

(a) $g(u) = (f(u))^{-1/2}$ if $f(s) > 0$ and $(\log f(s))'' \leq 0$ for $u_m \leq s \leq u_M$, i.e.

$$P = (f(u))^{-1/2}|\nabla u|^2 + \tfrac{2}{3}(f(u))^{3/2} = (f(u))^{-1/2}\{|\nabla u|^2 + \tfrac{2}{3}f^2(u)\}. \quad (5.33)$$

(b) If we choose $g(u) = e^{-\beta u}$, $\beta > 0$, then the equality sign holds in assumption (i), and (ii) is satisfied provided

$$(\log f(s))' \leq 2\beta \qquad \text{for} \quad u_m \leq s \leq u_M.$$

Then the function

$$P = e^{-\beta u}|\nabla u|^2 + \int_0^u e^{-\beta s} f(s)\, ds \tag{5.34}$$

assumes its maximum on ∂D.

5.3.2 Higher Dimensions

In this subsection, it is assumed that D is an N-dimensional domain with $N \geq 3$. A slightly different reasoning will prove

Theorem 5.2 Let u be a sufficiently smooth solution of (5.9) with $u_m \leq u \leq u_M$. If for $u_m \leq s \leq u_M$ the following assumptions are satisfied

(i) $g \neq 0$ and $(g^{-1})'' \leq 0$,
(ii) $(\log g)' \leq (2(1 - N)/(N + 2))(\log f)'$

then the function

$$P = g(u)|\nabla u|^2 + \frac{2}{N}\int_0^u f(s)g(s)\, ds$$

attains its maximum on ∂D.

Proof The combination of Lemma 5.4 and (5.14) implies that

$$\begin{aligned}\Delta P &- 2(\log g)' u_{,i} P_{,i} \\ &\geq |\nabla u|^4\left(g'' - 2\frac{g'^2}{g}\right) + |\nabla u|^2\left((h' - 2fg)'' \right. \\ &\quad \left. - \frac{g'}{g}(2h' - fg)\right) - f\left(h' - \frac{2}{N}fg\right).\end{aligned} \tag{5.35}$$

We choose h such that $h' = 2/N fg$. Then the coefficient of $|\nabla u|^2$ is nonnegative if (ii) is satisfied. Noting that

$$g'' - 2(g'^2/g) = -g^{-2}(g^{-1})'',$$

we see that the coefficient of $|\nabla u|^4$ is nonnegative also. Hence P satisfies

$$\Delta P - 2(\log g)' u_{,i} P_{,i} \geq 0, \qquad \text{in} \quad D,$$

from which the statement of Theorem 5.2 follows.

Examples (a) If $(f^{2(N-1/N+2)})'' \leq 0$ for $u_m \leq u \leq u_M$, we may take $g(u) = f^{2(1-N/N+2)}$, so that

$$P = f^{2(1-N)/(N+2)} |\nabla u|^2 + \frac{2}{N} \int_0^u (f(y))^{3N/(N+2)}\, dy. \tag{5.36}$$

(b) If $(g^{-1}(u))'' = 0$, i.e., $g = (\alpha u + \beta)^{-1}$, then assumption (ii) requires that for $u_m \leq u \leq u_M$ we have $\alpha u + \beta > 0$ and

$$f(u) \leq (\alpha u + \beta)^{2(N-1)/(N+2)}.$$

P then takes the form

$$P = \frac{|\nabla u|^2}{\alpha u + \beta} + \frac{2}{N} \int_0^u \frac{f(y)}{\alpha y + \beta}\, dy. \tag{5.37}$$

In Chapter 6, we shall see how Theorems 5.1 and 5.2 can be used to derive a number of bounds. In the next section, however, we shall derive conditions such that the corresponding functions P assume their maxima at a critical point of u.

5.4 THE MAXIMUM OF P AT A POINT WHERE $\nabla u = 0$

In this section we shall proceed as follows: We choose g and h such that the corresponding function P must have its maximum either on ∂D or at a point of D where $\nabla u = 0$. Should it occur that (under the appropriate conditions) the normal derivative $\partial P/\partial n$ is nonpositive on ∂D, then we would have reached a contradiction to the strong maximum principle. In such a case P must assume its maximum at a point where $\nabla u = 0$.

The analysis of $\partial P/\partial n$ depends on the boundary conditions that u satisfies. In all cases, however, it is convenient to choose $h(u)$ such that $h' = 2fg$. It will be assumed throughout that D possesses the "interior sphere property."

5.4.1 Dirichlet Boundary Conditions

Let u be a sufficiently smooth solution of (5.9) in a domain $D \subset E^N$ with zero-boundary data. Therefore

$$\frac{\partial P}{\partial n} = \frac{\partial}{\partial n}\left\{\left(\frac{\partial u}{\partial n}\right)^2 g(u) + h(u)\right\} = 2\frac{\partial u}{\partial n}\frac{\partial^2 u}{\partial n^2} g(0) + \left(\frac{\partial u}{\partial n}\right)^3 g'(0) + h'(0)\frac{\partial u}{\partial n}, \quad (5.38)$$

where the expression $(\partial^2 u)/(\partial n^2)$ is given as in (4.67) by

$$\frac{\partial^2 u}{\partial n^2} = u_{,ik} n^i n^k, \qquad \mathbf{n} = \text{outer normal on} \quad \partial D.$$

Let H be the average curvature of ∂D at a given point. Then since $u = 0$ on ∂D, according to (4.68), we have,

$$\frac{\partial^2 u}{\partial n^2} = \Delta u - (N-1)H\frac{\partial u}{\partial n} = -f(0) - (N-1)H\frac{\partial u}{\partial n}.$$

Using the fact that, also, $h'(0) = 2f(0)g(0)$ and $\partial u/\partial n = -|\nabla u|$, we arrive at

$$\frac{\partial P}{\partial n} = -|\nabla u|^2\{2g(0)(N-1)H + |\nabla u| g'(0)\}. \quad (5.39)$$

Note that if P assumes its maximum on ∂D, it must be where $|\nabla u| = \tau := \max_{\partial D} |\nabla u|$. Hence $\partial P/\partial n \leq 0$ on ∂D, provided that

$$2g(0)(N-1)H_0 + \tau g'(0) \geq 0. \quad (5.40)$$

Here $H \geq H_0 > -\infty$ on ∂D. In the following we combine (5.40) with Lemmas 5.1–5.3. We start with the special case where D is a plane domain.

Theorem 5.3 Let u be a sufficiently smooth solution of (5.9) in a plane domain D, vanishing on ∂D, and $u_m \leq u \leq u_M$. Suppose that for this range of u we have

(i) $g(u) > 0$, $f(u)g'(u) \leq 0$, $(\log g(u))'' \geq 0$.

(ii) Suppose that D is convex with curvature $k \geq k_0 \geq 0$ and $g'(0)\tau + 2k_0 g(0) > 0$.

Then the function

$$P = g(u)|\nabla u| + 2\int_0^u f(s)g(s)\,ds \quad (5.41)$$

attains its maximum where $\nabla u = 0$.

Proof Since $h' = 2gf$, it is easily seen that the quadratic form in (5.17) is positive semidefinite if (i) holds. Lemma 5.1 is therefore applicable. But under assumption (ii) $\partial P/\partial n \leq 0$ would follow. As pointed out at the beginning of Section 5.4, this implies that P must take its maximum where $\nabla u = 0$.

Remarks (a) It was shown in Sperb (1975) that if D is convex, then the solution of (5.9) with vanishing boundary values has only one critical point in D. In many important cases $f(u) \geq 0$, and thus $u \geq 0$ also. In this case u has exactly one maximum in D.

(b) A convenient choice is $g(u) = e^{-\beta u}$, $\beta > 0$. Then condition (ii) is satisfield if we choose $\beta = 2k_0/\tau$.

(c) In the event that ∂D is not convex, but $k_0 > -\infty$, one can, e.g., choose instead of (5.4)

$$P = g(u)|\nabla u|^2 + 2\int_0^u f(s)g(s)\,ds + \alpha u, \tag{5.42}$$

where α is sufficiently large.

(d) Of course τ is not known a priori, but it will be seen that Theorem 5.3 provides an upper bound for τ.

Let us now consider a domain in N dimensions with $N \geq 3$. Our next result will therefore be a combination of Lemma 5.2 and (5.40).

Theorem 5.4 Let u be a sufficiently smooth solution of (5.9) with zero-boundary data, such that $u_m \leq u \leq u_M$. Suppose that for this range of u we have

(i) $g(u) > 0$, $f(u)g'(u) \leq 0$, $(g^{-1/2}(u))'' \leq 0$, and
(ii) $2g(0)(N-1)H_0 + \tau g'(0) \geq 0$, where $H \geq H_0 > -\infty$.

Then the function

$$P = g(u)|\nabla u|^2 + 2\int_0^u f(s)g(s)\,ds$$

attains its maximum where $\nabla u = 0$.

Proof If (i) is satisfied, Lemma 5.2 is applicable. But condition (ii) again ensures that $\partial P/\partial n \leq 0$ on ∂D, from which the statement of Theorem 5.4 follows.

Remark (a) The following special case seems worthwhile mentioning:

Corollary 5.1 Suppose that u is a sufficiently smooth solution of (5.9) with zero-boundary data in a convex domain $D \subset E^N$, $N \geq 2$. Then the function

$$P = |\nabla u|^2 + 2\int_0^u f(s)\,ds \tag{5.43}$$

assumes its maximum where $\nabla u = 0$.

Proof The proof consists in noting that with $g \equiv 1$, $h = 2\int_0^u f(s)\,ds$, the assumptions of Theorems 5.3 and 5.4 are both satisfied if D is convex.

(b) The function P is optimal in the sense that $P \equiv \text{const}$ if D is a strip (for $N = 2$) or a slab ($N \geq 3$), when problem (5.9) reduces to (5.1).

(c) A simple choice is $g(u) = 1/(\alpha u + \beta)^2$, $\alpha > 0$, $\beta > 0$. Then if $f(u) > 0$, the assumptions of Theorem 5.4 are satisfied provided that $H_0 > 0$ and $\alpha = \beta(N-1)H_0/\tau$.

Finally, we may combine Lemma 5.3 with (5.40). This leads to a slightly different version of Theorem 5.4, namely

Theorem 5.5 Let u be a sufficiently smooth solution of (5.9) vanishing on ∂D. Suppose that for $u_m \leq u \leq u_M$ we have

(i) $g(u) > 0$, $f(u)g'(u) < 0$ if $u \neq 0$, $(g(u)^{-(N-2)/2(N-1)})'' \leq 0$, $N > 2$.
(ii) $2g(0)(N-1)H_0 + \tau g'(0) \geq 0$.

Then

$$P = g(u)|\nabla u|^2 + 2\int_0^u f(s)g(s)\,ds$$

assumes its maximum where $\nabla u = 0$.

Remark Note that

$$\frac{g'^2}{g} + \lambda g'' = \frac{\lambda^2}{\lambda+1}\frac{(g^{(\lambda+1)/\lambda})''}{g^{1/\lambda}}, \qquad \lambda \neq 1,$$

which allows us to write condition (iii) of Lemma 5.3 as

$$(g(u)^{-(N-2)/2(N-1)})'' \leq 0 \qquad \text{if} \quad N \geq 2.$$

This condition is weaker than

$$(g^{-1/2})'' < 0 \qquad \text{since} \quad \frac{N-2}{2(N-1)} = \frac{1}{2} - \frac{1}{2(N-1)}.$$

On the other hand, one needs the strict inequality $f(u)g'(u) < 0$ instead of $f(u)g'(u) \leq 0$.

5.4.2 Neumann Boundary Conditions

If $\partial u/\partial n = 0$ on ∂D, we can write on ∂D

$$P = |\tilde{\nabla} u|^2 g(u) + h(u),$$

where $\tilde{\nabla} u$ denotes the gradient in the $(N-1)$-dimensional hypersurface ∂D. Let the hypersurface ∂D be given in parametric form as

$$x^i = x^i(y^\alpha), \qquad \alpha = 1, \ldots, N-1, \qquad i = 1, \ldots, N.$$

Let $g_{\alpha\beta}$ be the metric tensor of ∂D, and denote covariant derivatives in the metric of ∂D (see Chapter 4) by a comma followed by a Greek subscript. Note that in this special case

$$g_{\alpha\beta} = \frac{\partial x^i}{\partial y^\alpha}\frac{\partial x^i}{\partial y^\beta}.$$

Now since

$$\frac{\partial u}{\partial n} = u_{,i}n^i = 0,$$

it follows that

$$\left(\frac{\partial u}{\partial n}\right)_{,\alpha} = u_{,i\alpha}n^i + u_{,i}n^i_{,\alpha} = 0.$$

The Weingarten formulas (4.11) give (note that **n** is the *exterior* normal)

$$n^i_{,\alpha} = h^\gamma_\alpha x^i_{,\gamma},$$

where h^γ_α is the mixed second fundamental tensor of ∂D.

Consider now $\partial P/\partial n$ at a point x on ∂D where P assumes a maximum. We choose our coordinate system such that at x the derivatives with Greek indices coincide with those with Latin indices for $i = 1, \ldots, N-1$. Since

$$\frac{\partial u}{\partial n} = u_{,N} = 0$$

we have

$$\begin{aligned}\frac{\partial P}{\partial n} &= 2g(u)u_{,\alpha}u_{,\alpha i}n^i = 2g(u)u_{,\alpha}u_{,i\alpha}n^i \\ &= -2g(u)u_{,\alpha}u_{,i}n^i_{,\alpha} \\ &= -2g(u)u_{,\alpha}u_{,i}h^\gamma_\alpha x^i_{,\gamma} \\ &= -2g(u)h^\gamma_\alpha u_{,\alpha}u_{,\gamma}.\end{aligned}$$

If we choose as the coordinate lines $y^\alpha = \text{const}$ the orthogonal lines of curvature, i.e., the eigendirections of h^γ_α with associated eigenvalues (sectional curvature) κ_α, we are then led to

$$\frac{\partial P}{\partial n} = -2g(u)\sum_{\alpha=1}^{N-1}\kappa_\alpha u_{,\alpha}u_{,\alpha}. \tag{5.44}$$

Note that for $N = 3$, $\kappa_1 + \kappa_2 = H$, whereas for $N = 2$, κ is the ordinary curvature of ∂D. The following result is now obvious.

Theorem 5.6 Let u be a sufficiently smooth solution of (5.3) in a convex domain $D \subset E^N$, $N \geq 2$, with $\partial u/\partial n = 0$ on ∂D. Suppose that assumption

(i) of Theorem 5.3 for $N = 2$ or assumptions (i) of Theorems 5.4 or 5.5 are satisfied. Then the function

$$P = g(u)|\nabla u|^2 + 2\int_0^u f(s)g(s)\,ds$$

assumes its maximum where $\nabla u = 0$.

Remarks (a) Again, a simple possibility is to take $g \equiv 1$, leading thus to the analog of Corollary 5.1:

Corollary 5.2 Let u be a sufficiently smooth solution of (5.9) with $\partial u/\partial n = 0$ on ∂D, and suppose that D is convex. Then

$$P = |\nabla u|^2 + 2\int_0^u f(s)\,ds$$

attains its maximum where $\nabla u = 0$.

(b) The calculations leading to (5.44) can be applied to an arbitrary smooth function v, which then yields the following result (as found by Payne, (1976) in the case of $N = 2, 3$).

Corollary 5.3 Let v be any $C^2(\bar{D})$ function satisfying $\partial v/\partial n = 0$ on a strictly convex portion Γ of ∂D. Then, if $|\nabla v|^2$ attains its maximum on Γ, it follows that $v \equiv$ const.

Proof According to (5.44), we have

$$\frac{\partial}{\partial n}|\nabla v|^2 = -2\sum_{\alpha=1}^{N-1} \kappa_\alpha v_{,\alpha} v_{,\alpha} < 0$$

by assumption. Hence $\max_D |\nabla v|^2 = 0$, i.e., $|\nabla v| \equiv 0$, so that $v \equiv$ const.

5.4.3 Robin- or Nonlinear Boundary Conditions

5.4.3.1 Plane Domains

Let D be a plane domain, and suppose that u satisfies

$$\frac{\partial u}{\partial n} + \sigma(u) = 0 \qquad \text{on} \quad \partial D, \qquad \sigma(u) \geq 0. \tag{5.45}$$

We now write on ∂D

$$P = \left(\left(\frac{\partial u}{\partial n}\right)^2 + \left(\frac{\partial u}{\partial s}\right)^2\right) g(u) + h(u) = \left(\sigma^2(u) + \left(\frac{\partial u}{\partial s}\right)^2\right) g(u) + h(u).$$

Then

$$\frac{\partial P}{\partial n} = 2\left(\frac{\partial u}{\partial n}\frac{\partial^2 u}{\partial n^2} + \frac{\partial u}{\partial s}\frac{\partial}{\partial n}\left(\frac{\partial u}{\partial s}\right)\right) g + |\nabla u|^2 g' \frac{\partial u}{\partial n} + h' \frac{\partial u}{\partial n},$$

and because of relations (4.2) and (4.3), as well as the boundary condition (5.45), the last equation can be cast in the form

$$\frac{\partial P}{\partial n} = 2g\left(-k|\nabla u|^2 - \frac{\partial^2 u}{\partial s^2}\frac{\partial u}{\partial n} - f\frac{\partial u}{\partial n} - \sigma'\left(\frac{\partial u}{\partial s}\right)^2\right) + |\nabla u|^2 g' \frac{\partial u}{\partial n} + h'\frac{\partial u}{\partial n}.$$

Choosing h such that $h' = 2gf$, we get

$$\frac{\partial P}{\partial s} = -2g\left(k|\nabla u|^2 + \frac{\partial^2 u}{\partial s^2}\frac{\partial u}{\partial n} + \sigma'\left(\frac{\partial u}{\partial s}\right)^2\right) + |\nabla u|^2 g' \frac{\partial u}{\partial n}. \tag{5.46}$$

Assume now that P attains its maximum at some point x of ∂D. There we must have

$$\frac{\partial P}{\partial s} = 2\left(\sigma\sigma' + \frac{\partial^2 u}{\partial s^2}\right)\frac{\partial u}{\partial s} g + |\nabla u|^2 g' \frac{\partial u}{\partial s} + h'\frac{\partial u}{\partial s} = 0. \tag{5.47}$$

If $\partial u/\partial s \neq 0$ at x, we conclude from (5.47) that

$$2\left(\sigma\sigma' + \frac{\partial^2 u}{\partial s^2}\right)g + |\nabla u|^2 g' + h' = 0, \tag{5.48}$$

i.e.,

$$2g\frac{\partial^2 u}{\partial s^2} = -(|\nabla u|^2 g' + h' + 2\sigma\sigma') = -(|\nabla u|^2 g' + 2fg + 2\sigma\sigma').$$

Inserting this into (5.46), we find

$$\frac{\partial P}{\partial n} = -2|\nabla u|^2\{g'\sigma + g(k + \sigma')\} - 2\sigma fg. \tag{5.49}$$

If $\partial u/\partial s = 0$ at x, we use the fact that $\partial^2 P/\partial s^2 \leq 0$ must hold at x. But if $\partial u/\partial s - 0$, then

$$\frac{\partial^2 P}{\partial s^2} = 2\left(\sigma\sigma' + \frac{\partial^2 u}{\partial s^2}\right)\frac{\partial^2 u}{\partial s^2} g + \sigma^2 g' \frac{\partial^2 u}{\partial s^2} + 2fg\frac{\partial^2 u}{\partial s^2}.$$

Suppose further that $\sigma' \geq 0$, $g > 0$ and that

$$2\sigma\sigma' g + \sigma^2 g' + 2fg \geq 0 \tag{5.50}$$

holds. Then $\partial^2 P/\partial s \leq 0$ implies that $\partial^2 u/\partial s^2 < 0$ at x. Hence if $\partial u/\partial s = 0$ at x, we have

$$\frac{\partial P}{\partial n} \leq -\sigma^2(2gk + \sigma g'). \tag{5.51}$$

In the next theorem, the necessary assumptions on f, g, σ are collected, ensuring that the corresponding function P satisfies a maximum principle and at the same time has the property that $\partial P/\partial n \leq 0$ if the maximum should occur on ∂D.

Theorem 5.7 Let u be a sufficiently smooth solution of (5.9) in a plane domain D, satisfying the boundary condition (5.45). Suppose the following assumptions hold

(i) For positive argument we have

$$g > 0, \qquad f \geq 0, \qquad \sigma > 0, \qquad g' \leq 0, \qquad (\log g)'' \geq 0.$$

(ii) The curvature k of ∂D is bounded below by $k_0 > 0$, and

$$\frac{g'}{g} + \frac{\sigma'}{\sigma} + \frac{k_0}{\sigma} \geq 0, \qquad 2\frac{\sigma'}{\sigma} + \frac{g'}{g} + 2\frac{f}{\sigma^2} \geq 0,$$

$$2\frac{g'}{g} + \frac{\sigma}{k_0} \geq 0.$$

Then the function

$$P = g(u)|\nabla u|^2 + 2\int_0^u f(y)g(y)\,dy$$

assumes its maximum at a point where $\nabla u = 0$.

Proof Under the conditions stated in assumptions (i) and (ii), the corresponding function P satisfies a maximum principle; however, if P would assume a maximum on ∂D, it would follow that $\partial P/\partial n \leq 0$ there. Hence, P must assume its maximum where $\nabla u = 0$.

An important special case is that in which $\sigma(u) = \alpha u$, $\alpha =$ const. > 0. Then the boundary condition (5.45) is often called the *Robin boundary condition*. If we make the simple choice

$$g(u) \equiv 1,$$

then Theorem 5.7 reduces to

Corollary 5.4 Let u be a sufficiently smooth solution of (5.9), satisfying Robin boundary conditions. Then if ∂D is convex, the function

$$P = |\nabla u|^2 + 2\int_0^u f(y)\,dy$$

assumes its maximum where $\nabla u = 0$.

5.4.3.2 Higher Dimensions

Unfortunately, the analysis used in Section 5.4.3.1 cannot be extended to higher dimensions. The reason is that not enough conditions can be given to determine the second-order derivatives of P that will appear in the discussion of the case in which P attains its maximum on ∂D. However, it seems very likely that the following is true.

Conjecture Let u be a sufficiently smooth solution of (5.9), satisfying the boundary condition (5.45). Suppose that $D \subset E^N$ is convex and $f \geq 0$, $\sigma' \geq 0$. Then the function

$$P = |\nabla u|^2 + 2 \int_0^u f(y)\, dy$$

assumes its maximum where $\nabla u = 0$.

It would be very interesting to know a different way to prove Theorem 5.7, in particular, a way that would also work for dimensions higher than two.

Chapter 6

APPLICATIONS

6.1 TORSION PROBLEM

Let u be the solution of

$$\Delta u = -2 \qquad \text{in} \quad D \tag{6.1}$$

vanishing on the boundary. Here D is a simply connected, finite plane domain. An important quantity in this problem is the maximum stress defined (up to a constant factor) as

$$\tau := \max_{\partial D} |\nabla u|. \tag{6.2}$$

Since

$$\begin{aligned} \Delta(|\nabla u|^2) = (u_{,i}u_{,i})_{,kk} = (2u_{,i}u_{,ik})_{,k} &= 2(u_{,ik}u_{,ik} + u_{,i}u_{,kki}) \\ &= 2u_{,ik}u_{,ik} \geq 0, \end{aligned}$$

it follows that indeed $\max_D |\nabla u| = \max_{\partial D} |\nabla u|$.

We shall see that Theorems 5.1 and 5.3 can be applied to gain valuable information on τ. First we choose, in Theorem 5.1, $g(u) \equiv 1$ so that $P = |\nabla u|^2 + 2u$. Theorem 5.1 then tells us that P must attain its maximum on ∂D. We can express this result in the form of the inequality

$$|\nabla u|^2 + 2u \leq \tau^2 \qquad \text{in} \quad D. \tag{6.3}$$

Furthermore, at any point of ∂D where P assumes its maximum we must have

$$\frac{\partial P}{\partial n} = \frac{\partial}{\partial n}(|\nabla u|^2 + 2u) > 0,$$

unless $P \equiv$ const. But

$$\frac{\partial}{\partial n}(|\nabla u|^2 + 2u) = 2\frac{\partial u}{\partial n}\frac{\partial^2 u}{\partial n^2} + 2\frac{\partial u}{\partial n} = 2\frac{\partial u}{\partial n}\left(-2 - k\frac{\partial u}{\partial n}\right) + 2\frac{\partial u}{\partial n}$$
$$= -2\frac{\partial u}{\partial n} - 2k\left(\frac{\partial u}{\partial n}\right)^2 > 0,$$

which we may write as

$$k\tau < 1. \tag{6.4}$$

Here k denotes the curvature of ∂D at a point where $|\nabla u| = \tau$ on ∂D. If $k \geq k_0 > 0$, it thus follows that

$$\tau \leq \frac{1}{k_0}. \tag{6.5}$$

Remarks (a) It is easy to check that the equality sign in (6.5) holds when D is a circle, in which case P becomes a constant.

(b) It was pointed out by Bandle (1979) that inequality (6.5) also follows from a purely geometric result combined with the monotonicity of u with respect to D. However, inequality (6.3) contains other information as well, as illustrated in the following.

Let $S := \int_D |\nabla u|^2\,dx$ be the torsional rigidity of D. If we integrate (6.3) over D, observing that we have $S = 2\int_D u\,dx$ because of Green's identity, we find

$$\tau^2 \geq 2S/A, \qquad A = \text{area of} \quad D. \tag{6.6}$$

A combination of (6.5) and (6.6) then yields

$$k(x) \leq \sqrt{A/2S}, \tag{6.7}$$

where $k(x)$ is the curvature at the point x where $|\nabla u| = \tau$.

Example If D is an ellipse of semiaxes a, b, one has $A = ab\pi$, $S = \pi a^3 b^3/(a^2 + b^2)$, so that (6.7) tells us that $k(x) \leq \frac{1}{2}\sqrt{(1/a^2) + (1/b^2)}$ must hold. We can also evaluate (6.3) at a point where u assumes its maximum u_M to find

$$\tau^2 \geq 2u_M. \tag{6.8}$$

A possible lower bound for u_M is the corresponding value for the largest inscribed disk of radius ρ, for which $u_M = \frac{1}{2}\rho^2$, and therefore

$$\tau \geq \rho. \tag{6.9}$$

Finally, the combination of (6.5) and (6.6) yields

$$S \leq \tfrac{1}{2}A/k_0^2. \tag{6.10}$$

Of course this last bound will in general be crude. Nevertheless, (6.10) and all the other bounds derived so far are optimal in the sense that equality holds throughout if D is a disk.

Let us now turn to applications of Theorem 5.3. The simplest version is to choose $g(u) \equiv 1$ again. Then Theorem 5.3 states that the function $P = |\nabla u|^2 + 4u$ assumes its maximum where $\nabla u = 0$ if D is convex. Or, to put it differently, one has

$$|\nabla u|^2 + 4u \leq 4u_M. \tag{6.11}$$

This inequality can now be used in a number of ways. For instance on ∂D we get

$$\tau^2 \leq 4u_M. \tag{6.12}$$

As an upper bound for u_M one may take

(a) the smallest circumscribed circle; then

$$u_M \leq \frac{|D|^2}{2}, \qquad \text{with} \quad |D| = \text{diameter of } D.$$

Another possibility is

(b) the smallest strip of width $2d$ containing D. Then a simple computation gives $u_M \leq d^2$.

In the following we shall see that it is possible to derive from (6.11) a sharper bound than the ones given in (a) and (b). To this end, let M be the point where $u = u_M$ (there is only one such point if D is convex (see Sperb, (1975) and Q a point on ∂D nearest to M. Let r measure the distance from M along the ray connecting M and Q. Certainly $-du/dr \leq |\nabla u|$ and therefore by (6.11)

$$\frac{du}{dr} \leq 2\sqrt{u_M - u},$$

and hence

$$\int_0^{u_M} \frac{du}{2\sqrt{u_M - u}} \leq \int_M^Q dr =: \overline{MQ}.$$

The best value for $\overline{MQ}$ is again $\overline{MQ} = \rho$, the radius of the largest circle inscribed in D. We have thus found that

$$u_M \leq \rho^2, \tag{6.13}$$

which is in general sharper than the bounds given under (a) and (b) (take, e.g., for D a triangle!). Yet another way to use (6.11) is as follows: we integrate our last integral not from M to Q, but from M to a variable point $R \subset D$ and set $MR = r$. Then

$$u_M - u(r) \leq r^2,$$

which we now integrate over D to find

$$Au_M \leq I_0 + \tfrac{1}{2}S, \tag{6.14}$$

where I_0 is the polar moment of inertia of D with respect to M. If D has two axes of symmetry, one knows that M must be at the center and I_0 can be calculated. For comparison note that for a circle of radius R (6.13) gives $u_M \leq R^2$, whereas (6.14) yields $u_M \leq \frac{3}{4}R^2$ and the correct value is $u_M = \frac{1}{2}R^2$. Inequality (6.13), however, is exact if D is a strip, so that we can expect that (6.13) is good for a "long" domain.

An integration of (6.11) over D shows that

$$3S \leq 4Au_M. \tag{6.15}$$

On the other hand, we may write (6.11) on ∂D as

$$-\frac{\partial u}{\partial n} \leq 2\sqrt{u_M}$$

and then integrate over ∂D to find that

$$u_M \geq (A^2/L^2). \tag{6.16}$$

Since inequality (6.11) becomes an equality only if D is a strip, it will be somewhat crude, e.g., if D is a circle. But Theorem 5.3 allows to improve considerably some of the inequalities already given.

It is not hard to check that the function

$$P = |\nabla u|^2 e^{-\beta u} + 4\int_0^u e^{-\beta y}\,dy, \tag{6.17}$$

where $\beta = 2k_0/\tau$ and $k \geq k_0 > 0$, satisfies the assumptions of Theorem 5.3 and therefore attains its maximum where $u = u_M$. Instead of (6.12) we then have

$$\tau \leq \frac{2}{k_0}(1 - e^{(-2k_0u_M)/\tau}).$$

Setting $\bar{x} = 2k_0u_M/\tau$, we thus see that

$$\tau < \frac{2k_0u_M}{\bar{x}}, \tag{6.18}$$

where $x = \bar{x}$ is the positive solution of

$$x(1 - e^{-x}) = k_0^2u_M. \tag{6.19}$$

We determine an upper bound for u_M analogously as before. The inequality corresponding to (6.12), with P given by (6.17), is

$$|\nabla u|^2 \leq 4 \int_u^{u_M} e^{-\beta(v-u)}\,dv = \frac{4}{\beta}(1 - e^{-\beta(u_M - u)}), \tag{6.20}$$

so that it now follows that

$$\int_0^{u_M} (1 - e^{-\beta(u_M - u)})^{-1/2}\,du \leq \frac{2}{\sqrt{\beta}}\rho. \tag{6.21}$$

The integral in (6.21) can be calculated analytically, and after some steps one finds that

$$\frac{\sin^2 x_0}{(1 + (1 - \sin^2 x_0)^{1/2})^2} \geq e^{-2\sqrt{\beta}\rho}, \qquad \sin^2 x_0 = e^{-\beta u_M}. \tag{6.22}$$

Since the left side in (6.22) is increasing with $\sin^2 x_0$ we may use (6.18) to find that

$$\tau \leq 2\hat{x}^2/k_0, \tag{6.23}$$

where $\hat{x}$ is the positive solution of

$$x \operatorname{artanh} x = k_0\rho. \tag{6.24}$$

A simple series expansion in (6.24) gives the cruder result

$$\tau \leq ((9 + 12k_0\rho)^{1/2} - 3)/k_0, \tag{6.25}$$

which is nevertheless better than the combination of (6.12) and (6.13), giving

$$\tau \leq 2\rho, \tag{6.26}$$

as it is not hard to check.

In order to give somc idea what the bounds yield, we list in the accompanying tabulation, a few values for the case in which D is an ellipse of semi-axes a, b,. A number of additional inequalities could be derived along similar lines. The interested reader may refer to the papers of Payne et al. (1977) and Payne (1968).

a/b	Exact value of τ/a	(6.23)	(6.26)
1	1	1.394	2
2	0.8	0.918	1
3	0.6	0.642	0.667
4	0.471	0.489	0.5
5	0.385	0.394	0.4

6.2 $\Delta u + \lambda u = 0$

6.2.1 A Lower Bound for the First Eigenvalue for Robin Boundary Conditions

Let u be the first, strictly positive eigenfunction of

$$\Delta u + \lambda u = 0 \qquad \text{in} \quad D, \tag{6.27}$$

$$\frac{\partial u}{\partial n} + \alpha u = 0 \qquad \text{on} \quad \partial D, \tag{6.28}$$

where $\alpha > 0$, $\alpha = \text{const}$, D is a plane domain, and $\lambda = \lambda_1(\alpha) > 0$ is the first eigenvalue in problem (6.27), (6.28).

As a consequence of Theorem 5.1, we shall prove the inequality

$$\lambda_1(\alpha) > \alpha\left(2k_0 - \frac{\alpha}{2}\right), \tag{6.29}$$

where $0 < \alpha \leq k_0 \leq k$ is the curvature of ∂D.

In order to prove (6.29), we note first that $g(u) = u^{-1/2}$ satisfies assumptions (i) and (ii) of Theorem 5.1, and therefore the function

$$P := \frac{|\nabla u|^2}{u^{1/2}} + \lambda \int_0^u v^{1/2}\, dv = u^{-1/2}\left(|\nabla u|^2 + \frac{2}{3}\lambda u^2\right)$$

must assume its maximum at some point x of ∂D. The strong maximum principle implies that at x we must have (for nonconstant P)

$$\frac{\partial P}{\partial n} > 0.$$

The following calculations resemble closely those in Section 5.4.3. As in Section 5.4.3, we first write

$$P = u^{-1/2}\left\{\left(\frac{\partial u}{\partial n}\right)^2 + \left(\frac{\partial u}{\partial s}\right)^2 + \frac{2}{3}\lambda u^2\right\},$$

so that

$$\frac{\partial P}{\partial n} = -\frac{1}{2}\left\{\frac{\partial u}{\partial n}\left(\frac{\partial u}{\partial n}\right)^2 + \left(\frac{\partial u}{\partial s}\right)^2 + \frac{2}{3}\lambda u^2\right\}u^{-3/2}$$
$$+ u^{-1/2}\left\{2\frac{\partial u}{\partial n}\frac{\partial^2 u}{\partial n^2} + 2\frac{\partial u}{\partial n}\frac{\partial}{\partial n}\left(\frac{\partial u}{\partial s}\right) + \frac{4}{3}\lambda u\frac{\partial u}{\partial n}\right\}.$$

Again we make use of the relations

$$\frac{\partial}{\partial n}\left(\frac{\partial u}{\partial s}\right) = \frac{\partial}{\partial s}\left(\frac{\partial u}{\partial n}\right) - k\frac{\partial u}{\partial s} = -(\alpha + k)\frac{\partial u}{\partial s}$$

and

$$\Delta u = \frac{\partial^2 u}{\partial n^2} + k\frac{\partial u}{\partial n} + \frac{\partial^2 u}{\partial s^2} = -\lambda u$$

on ∂D. This allows us to write, after some simplifications,

$$u^{1/2}\frac{\partial P}{\partial n} = u^2\left(\lambda\alpha + \frac{\alpha^3}{2} - 2k\alpha^2\right) + 2\alpha u\frac{\partial^2 u}{\partial s^2} - \left(\frac{\partial u}{\partial s}\right)^2\left(\frac{3}{2}\alpha + 2k\right) > 0. \quad (6.30)$$

At x we have also

$$\begin{aligned}\frac{\partial P}{\partial s} &= \frac{\partial}{\partial s}\left\{\alpha^2 u^{3/2} + u^{1/2}\left(\frac{\partial u}{\partial s}\right)^2 + \frac{2}{3}\lambda u^{3/2}\right\}\\ &= \frac{3}{2}\alpha^2 u^{1/2}\frac{\partial u}{\partial s} - \frac{1}{2}u^{-3/2}\left(\frac{\partial u}{\partial s}\right)^3 + 2u^{1/2}\frac{\partial u}{\partial s}\frac{\partial^2 u}{\partial s^2} + \lambda u^{1/2}\frac{\partial u}{\partial s} = 0. \quad (6.31)\end{aligned}$$

If $\partial u/\partial s \neq 0$ at x, we conclude from (6.31) that

$$\left(\lambda + \frac{3}{2}\alpha^2\right)u^{1/2} + 2u^{-1/2}\frac{\partial^2 u}{\partial s^2} - \frac{1}{2}u^{-3/2}\left(\frac{\partial u}{\partial s}\right)^2 = 0,$$

i.e.,

$$2\alpha u\frac{\partial^2 u}{\partial s^2} = \frac{1}{2}\left(\frac{\partial u}{\partial s}\right)^2 - \alpha\left(\lambda + \frac{3}{2}\alpha^2\right)u^2,$$

which we may insert into (6.30) to find

$$\frac{\partial P}{\partial n} = u^2\left(\lambda\alpha + \frac{\alpha^3}{2} - 2k\alpha^2\right) - \left(\frac{\partial u}{\partial s}\right)^2(\alpha + 2k) - \alpha\left(\lambda + \frac{3}{2}\alpha^2\right)u^2 > 0. \quad (6.32)$$

In the case that $\partial u/\partial s = 0$ at x, we have

$$P = (\alpha^2 + \tfrac{2}{3}\lambda)u^{3/2}$$

there, which clearly implies that the maximum of P on ∂D must be at a point where u itself is a maximum, and therefore

$$\frac{\partial^2 u}{\partial s^2} \leq 0 \qquad \text{at} \quad x.$$

Another look at (6.30) and (6.32) then tells us that the inequality

$$\lambda\alpha + \frac{\alpha^3}{2} - 2k\alpha^2 \geq 0$$

must hold, i.e., inequality (6.29).

Remarks (a) Obviously (6.29) is only interesting for strictly convex domains ($k \geq k_0 > 0$) and $0 < \alpha \leq 4k_0$.

(b) Inequality (6.29) is exact asymptotically if D is a disk in the sense that

$$\lim_{\alpha\to 0} \frac{\lambda_1(\alpha)}{\alpha} = \frac{L}{A}$$

for an arbitrary domain, and thus for a circle

$$\lim_{\alpha\to 0} \frac{\lambda_1(\alpha)}{\alpha} = 2k$$

as indicated by (6.29). Philippin (1978) has proven the weaker inequality

$$\lambda_1(\alpha) \geq \alpha(k_0 - \alpha).$$

Different types of bounds for $\lambda_1(\alpha)$ can be found in Sperb (1972).

6.2.2 A Bound for the "Efficiency Ratio"

The application of Theorem 5.4, Corollary 5.1, described in this section was given by Payne and Stakgold (1972). The interested reader is also referred to the latter paper for more details on the physical background. In the context of nuclear reactors, the equation

$$\Delta u + \lambda u = 0 \qquad \text{in} \quad D, \tag{6.33}$$

where u now vanishes on ∂D, is also of some interest. Specifically, if λ_1 is the first (positive) eigenvalue and u the corresponding eigenfunction, the "efficiency ratio" E is defined as

$$E := \frac{\int_D u\,dx}{u_M A}, \tag{6.34}$$

where $u_M = \max_{x\in D} u(x)$, and A is the volume (or area) of D.

In order to obtain an estimate for E we use Corollary 5.1, which tells us that if D is convex, the function

$$P = |\nabla u|^2 + \lambda_1 u^2$$

assumes its maximum where $u = u_M$, i.e.,

$$|\nabla u|^2 + \lambda_1 u^2 \leq \lambda_1 u_M^2. \tag{6.35}$$

On ∂D, (6.35) simplifies to

$$-\frac{\partial u}{\partial n} \leq \sqrt{\lambda_1}\, u_M,$$

which can be integrated over ∂D to give

$$\lambda_1 \int_D u\, dx \leq \lambda_1 L u_M, \qquad L = \text{surface area (or length) of } \partial D.$$

As a first result, we therefore find that for convex domains D

$$E := \left(\int_D u\, dx\right)\Big/ u_M A \leq L/\sqrt{\lambda_1} A. \tag{6.36}$$

As before we can also give a lower bound for λ_1 (which is needed in (6.36)). Assume that $u = u_M$ at M and that R is a point of ∂D. Then as before we use the fact that

$$-\frac{du}{dr} \leq |\nabla u| < \sqrt{\lambda_1}\sqrt{u_M^2 - u^2},$$

i.e.,

$$\int_0^{u_M} \frac{du}{\sqrt{u_M^2 - u^2}} \leq \sqrt{\lambda_1} \int_M^R dr = \lambda_1 \overline{RM}.$$

Again the optimal value for $\overline{RM}$ is ρ = radius of the largest inscribed sphere (or circle), and we have

$$\lambda_1 \geq \pi^2/4\rho^2, \tag{6.37}$$

an inequality found by Hersch (1960), using different arguments. Insertion of (6.37) into (6.36) finally leads to

$$E \leq \frac{2}{\pi}\frac{\rho L}{A} \qquad (D \text{ convex}). \tag{6.38}$$

As we will see in Section 6.4, inequality (6.38) can still be improved, and we will show as in Payne and Stakgold (1972) that actually

$$E \leq 2/\pi. \tag{6.39}$$

The equality sign in (6.38) and (6.39) holds in the limiting case in which (in two dimensions) D approaches a strip. On the other hand, since certainly

$$\rho L > A,$$

inequality (6.39) is sharper than (6.38). For plane regions, however, Theorem 5.3 allows us to improve (6.39) considerably in some cases. Following

Schaefer and Sperb (1976), we take

$$P = |\nabla u| e^{-\beta u} + 2\lambda \int_0^u e^{-\beta v} v \, dv,$$

where $\beta = \partial k_0/\sigma$ and $\sigma = \max_{\partial D} |\nabla u|$. Here D is assumed to be strictly convex with $k \geq k_0 > 0$.

On ∂D one thus has

$$|\nabla u|^2 \leq 2\lambda \int_0^{u_M} e^{-\beta v} v \, dv,$$

and after some manipulation a first result is that

$$\sigma/u_M \leq 2k_0/x, \tag{6.40}$$

where $\bar{x}$ is the positive solution of

$$e^{-x}(1 + x) = 1 - 2k_0^2/\lambda_1. \tag{6.41}$$

Green's identity and the bound (6.40) finally lead to

$$E \leq \frac{2L}{\lambda A} \frac{k_0}{x}. \tag{6.42}$$

For D a disk, (6.42) gives $E \leq 0.565$, whereas (6.39) tells us that $E \leq 0.636$.

Remarks (a) For nonconvex regions the functions P have to be modified. But it is still possible to obtain bounds for E along the same lines, as shown in Payne and Stakgold, (1972).

(b) If the boundary conditions are changed into (6.28) instead of Dirichlet boundary conditions, one can derive similar bounds for E (see Schaefer and Sperb (1976).

(c) The function P chosen to prove (6.42) can also be used to derive a lower bound for λ_1. The calculations, however, become rather involved and are therefore omitted here.

6.2.3 Improvement of an Inequality of Payne and Weinberger for Symmetric Plane Regions

Let μ_2 be the first positive eigenvalue in the "free membrane problem"

$$\Delta u + \mu u = 0 \quad \text{in} \quad D \subset E^2, \tag{6.43}$$

$$\frac{\partial u}{\partial n} = 0 \quad \text{on} \quad \partial D. \tag{6.44}$$

The first eigenvalue μ_1 is zero and the associated eigenfunction is just a constant. The second eigenfunction u_2 must change sign, and by the famous nodal line theorem of Courant and Hilbert (1953), it follows that D is divided into two subdomains D^+ and D^- such that $u_2 > 0$ in D^+ and $u < 0$ in D^-. It can be shown (see Payne, (1967) that u_2 cannot have a closed nodal line (i.e., the curve where $u_2 = 0$) in D. On the other hand, if D has two axes of symmetry, the same is true for the corresponding eigenfunction u_2. Hence if D has two axes of symmetry, the nodal line of u_2 must contain one of the axes.

According to Corollary 5.2, for convex D the function

$$P = |\nabla u_2|^2 + \mu_2 u_2^2$$

must assume its maximum where $|u_2|$ assumes its maximum u_M. Suppose this is at the point M and Z is the center of symmetry, i.e., the point of intersection of the two axes.

Note that the point M must lie on one of the axes. As before we have

$$\int_0^{u_M} \frac{du}{\sqrt{u_M^2 - u^2}} \leq \mu_2 \overline{MZ} \leq \mu_2 \frac{|a|}{2},$$

where $|a|$ is the length of longer axis of D, so that

$$\mu_2 \geq \pi^2/|a|^2 \tag{6.45}$$

if D is convex and symmetric.

Payne and Weinberger (1960) showed by entirely different methods that for any convex plane domain one has

$$\mu_2 \geq \pi^2/|D|^2,$$

where $|D|$ is the diameter of D, which is cruder than (6.45), but no symmetry assumption is needed for the validity of their inequality. On the other hand, (6.45) can be extended to higher dimensions and also to domains on a Riemannian manifold as will be shown in Chapter 8. Note that (6.45) is exact for any rectangle.

Remark The convexity assumption is essential since μ_2 can become arbitrarily small for a domain of a barbell shape as shown in Figure 6.1. Let

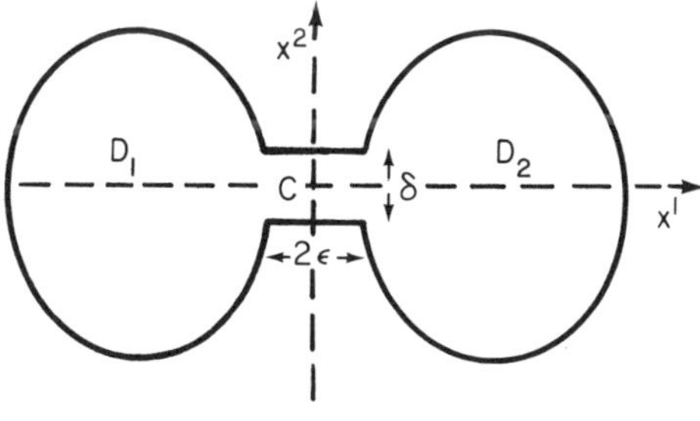

Figure 6.1

D be composed of two disks D_1, D_2 of area A connected by a channel C of length 2ε and width δ. We shall see that $\mu_2(D)$ can become as small as we wish if the channel C is chosen appropriately. The Raleigh-Principle (see, e.g., Courant and Hilbert (1953) states that

$$\mu_2 = \min_{\int_D v\,dx=0} \frac{\int_D |\nabla v|^2\,dx}{\int_D v^2\,dx} =: R[v].$$

We now choose

$$v(x^1, x^2) = \begin{cases} -\varepsilon, & x^1 \le -\varepsilon \\ x^1, & -\varepsilon < x^1 < \varepsilon \\ \varepsilon, & x^1 \ge \varepsilon. \end{cases}$$

Then the side condition is obviously satisfied and

$$R[v] = \frac{2\varepsilon\delta}{2\varepsilon^2 A + \frac{2}{3}\varepsilon^2\delta} = \frac{\delta}{\varepsilon A + \frac{1}{3}\varepsilon\delta},$$

i.e.,

$$\mu_2 \le \frac{\delta}{\varepsilon A + \frac{1}{3}\varepsilon\delta},$$

from which it is clear that μ_2 becomes small if δ is chosen small.

6.3 A LOWER BOUND FOR THE FIRST STEKLOFF EIGENVALUE

The Stekloff problem in its original form is the eigenvalue problem

$$\Delta h = 0 \quad \text{in} \quad D, \tag{6.46}$$

$$\frac{\partial h}{\partial n} = ph \quad \text{on} \quad \partial D, \tag{6.47}$$

where D is a plane domain and the first nonzero eigenvalue p_2 ($p_1 = 0$, $h_1 = \text{const}$) is of special interest. Note that Theorem 5.1 also contains the well-known result that

$$P := |\nabla h|^2 \tag{6.48}$$

must assume its maximum on ∂D. A similar reasoning as the one applied in Section 6.2.1 allows us to prove the inequality (Payne, 1970)

$$p_2 \ge k_0, \tag{6.49}$$

where $k \ge k_0 > 0$ is assumed, i.e., D is strictly convex.

For the proof we again use the fact that according to the strong maximum principle we must have at a point x of ∂D where P is a maximum

$$\frac{\partial P}{\partial n} = 2\frac{\partial h}{\partial n}\frac{\partial^2 h}{\partial n^2} + 2\frac{\partial h}{\partial s}\frac{\partial}{\partial n}\left(\frac{\partial h}{\partial s}\right) > 0 \tag{6.50}$$

if $P \not\equiv$ const. Again we use the fact that on ∂D

$$\Delta h = \frac{\partial^2 h}{\partial n^2} + k\frac{\partial h}{\partial n} + \frac{\partial^2 h}{\partial s^2} = 0,$$

and

$$\frac{\partial}{\partial n}\left(\frac{\partial h}{\partial s}\right) = \frac{\partial}{\partial s}\left(\frac{\partial h}{\partial n}\right) - k\frac{\partial h}{\partial s} = (p_2 - k)\frac{\partial h}{\partial s}.$$

Insertion into (6.50) yields

$$\frac{\partial P}{\partial n} = -2k\left(\frac{\partial h}{\partial n}\right)^2 - 2\frac{\partial h}{\partial n}\frac{\partial^2 h}{\partial s^2} + 2(p_2 - k)\left(\frac{\partial h}{\partial s}\right)^2 > 0. \tag{6.51}$$

Furthermore, at x we have

$$\frac{\partial P}{\partial s} = 2\frac{\partial h}{\partial n}\frac{\partial}{\partial s}\left(\frac{\partial h}{\partial n}\right) + 2\frac{\partial h}{\partial s}\frac{\partial^2 h}{\partial s^2} = 2\left(p_2\frac{\partial h}{\partial n} + \frac{\partial^2 h}{\partial s^2}\right)\frac{\partial h}{\partial s} = 0. \tag{6.52}$$

Hence, if $\partial h/\partial s \neq 0$ at x, it follows that

$$\frac{\partial^2 h}{\partial s^2} = -p_2\frac{\partial h}{\partial n},$$

and (6.50) can be rewritten as

$$(p_2 - k)|\nabla h|^2 > 0, \tag{6.53}$$

which implies (6.49). If $\partial h/\partial s = 0$ at x, then we may use the fact that $\partial^2 P/\partial s^2 \leq 0$ at x also. But since $\partial h/\partial s = 0$,

$$\frac{\partial^2 P}{\partial s^2} = 2\left(p_2^2 h + \frac{\partial^2 h}{\partial s^2}\right)\frac{\partial^2 h}{\partial s^2} \leq 0,$$

i.e., either at x we have

(a) $$\frac{\partial^2 h}{\partial s^2} \leq 0 \quad \text{and} \quad \frac{\partial^2 h}{\partial s^2} + p_2^2 h \geq 0$$

or

(b) $$\frac{\partial^2 h}{\partial s^2} \geq 0 \quad \text{and} \quad \frac{\partial^2 h}{\partial s^2} + p_2^2 h \leq 0$$

In case (a) we thus have $p_2^2 h \geq -\partial^2 h/\partial s^2 \geq 0$, i.e., $h \geq 0$ at x, and since $P = p_2 h^2$ there, we know that h is itself a maximal at x. In case (b) we have $0 \leq \partial^2 h/\partial s^2 \leq -p_2^2 h$, implying that h has a minimum at x.

Going back to (6.51), we note that the expression

$$-2\frac{\partial h}{\partial n}\frac{\partial^2 h}{\partial s^2} = -2p_2 h\frac{\partial^2 h}{\partial s^2}$$

is positive (nonnegative) in any case at x. Consequently we have

$$-2\frac{\partial h}{\partial n}\frac{\partial^2 h}{\partial s^2} \geq 2p_2\frac{\partial h}{\partial n},$$

and therefore (6.51) implies (6.53) again. This completes the proof.

Remarks (a) It is not hard to check that (6.49) becomes an equality if D is a disk, when P as given by (6.48) is a constant, since $h(x^1, x^2) = x^1$ (or x^2).

(b) Actually, Theorem 5.1 implies the more general result that for any $\beta > 0$ the function

$$P = e^{-\beta h}|\nabla h|^2$$

attains its maximum on ∂D. However, the calculation shows that the optimal value of β in view of the application to problem (6.46), (6.47) is $\beta = 0$.

(c) With reasoning analogous to that above, it is possible to determine bounds for $\max_{\partial D}|\nabla h|^2$ in terms of the boundary values of h (and their derivatives) for an arbitrary harmonic function h.

6.4 SOME BOUNDS IN A NONLINEAR EIGENVALUE PROBLEM

Many important problems in physics, chemistry, or biology lead to the nonlinear problem

$$\Delta u + \lambda f(u) = 0 \qquad \text{in} \quad D \tag{6.54}$$

with zero boundary data. Here λ is a positive parameter, and only positive solutions are of practical interest most of the time. In addition, the "force term" $f(u)$ is in many instances positive as well, i.e., $f(s) > 0$ for $s > 0$. As always, we shall only consider sufficiently smooth solutions, which in our case will mostly mean $C^3(D) \cap C^2(\bar{D})$. The regularity results cited in Chapter 3 ensure that if f and ∂D are smooth, so are the solutions of (6.54).

6.4.1 Bounds for the Critical Values

The set of values λ, for which (6.54) with u vanishing on ∂D has a positive solution, is called the spectrum. As we shall see, the spectrum is an interval (λ_*, λ^*), where $0 \le \lambda_* < \lambda^* \le \infty$ and the endpoints λ_*, λ^* may or may not belong to the spectrum. The linear case $f(u) = u$ is exceptional; then $\lambda_* = \lambda^*$, i.e., the spectrum consists only of the first positive eigenvalue λ_1 of (6.33). Another important distinction to be made is between the "forced case" $f(0) > 0$ and the "unforced case" $f(0) = 0$. In the former case $\lambda_* = 0$, whereas if $f(0) = 0$, more information on the behavior of $f(s)$ near $s = 0$ is required.

Our first result is an immediate consequence of Theorem 5.4 (Corollary 5.1) and gives a lower bound for λ_* in the unforced case.

Theorem 6.1 Let D be a convex domain of E^N and suppose that

(a) $\lim_{s\to 0} (s/f(s)) = l_0 > 0$, $\lim_{s\to\infty} (s/f(s)) = l_\infty > 0$, f increasing for $s > 0$.

(b) $\inf_{s>0} H(s) = H_0$ with $H(s) := \int_0^S (F(s) - F(t))^{-1/2}\, dt$, $dF/ds = f(s)$.

Then

$$\lambda_* \ge H_0^2/2\rho^2 > 0,$$

where ρ is the radius of the largest ball contained in D.

Proof According to Corollary 5.1, we have

$$|\nabla u|^2 + 2\lambda F(u) \le 2\lambda F(u_M), \qquad u_M = \max_{x\in D} u(x).$$

With the notation of Section 6.1, we therefore have

$$-\frac{du}{dr} \le (2\lambda)^{1/2}(F(u_M) - F(u))^{1/2},$$

and by integration, as before,

$$H(u_M) := \int_0^{u_M} \frac{du}{(F(u_M) - F(u))^{1/2}} \le (2\lambda)^{1/2}\rho.$$

In particular, it follows that

$$H_0 := \inf_{s>0} H(s) \le (2\lambda)^{1/2}\rho.$$

Since f is increasing, we have for $s > t$

$$F(s) - F(t) \le f(s)(s - t),$$

so that

$$H(s) \ge f(s)^{-1/2} \int_0^s \frac{dt}{(s - t)^{1/2}} = 2\left(\frac{s}{f(s)}\right)^{1/2}.$$

Because of our assumptions in (a), we see that

$$H_0 > 0,$$

and clearly we must have

$$\lambda \geq \lambda_* \geq H_0^2/2\rho^2 > 0. \tag{6.55}$$

This proves Theorem 6.1.

So far we know only that if λ is too small, problem (6.54) has *no* positive solution for the nonlinearity described in Theorem 6.1. At this point let us recall inequality (3.16) which states that

$$\lambda_1 \inf_{s>0} \frac{s}{f(s)} \leq \lambda \leq \lambda_1 \sup_{s>0} \frac{s}{f(s)}$$

must hold, with λ_1 = first eigenvalue of (6.33). In other words, we also have

$$\lambda_* \geq \lambda_1 \inf_{s>0} \frac{s}{f(s)} \tag{6.56}$$

and

$$\lambda^* \leq \lambda_1 \sup_{s>0} \frac{s}{f(s)}. \tag{6.57}$$

Note that $H_0^2 > 4 \inf_{s>0} (s/f(s))$, but by (6.37) $\lambda_1 \geq \pi^2/4\rho^2$, so that the bounds (6.55) and (6.56) cannot be compared.

Inequality (6.55) is optimal in the sense that we have equality, if problem (6.54) can be reduced to a one-dimensional problem and f is arbitrary. In contrast to (6.54), (6.56) is optimal in the sense that we have equality if $f(s) = s$ for any domain D.

Inequality (6.57) tells us that $\lambda^* < \infty$ if $\sup_{s>0} (s/f(s)) < \infty$. It is a consequence of a general result of Amann and Laetsch (1976) that this problem has at least one positive solution for

$$\lambda \in (\lambda_*, \lambda^*) \qquad \text{if} \quad \lim_{s\to\infty} \frac{s}{f(s)} = l_\infty < \infty$$

and

$$\lim_{s\to 0} \frac{f}{f(s)} = l_0 = \frac{1}{f'(0)} < \infty.$$

In addition $\lambda_* \geq \lambda_1 l_\infty$ and $\lambda^* \leq \lambda_1 l_0$ if f is convex and strictly increasing. Summarizing, we can thus state that if $g(s) := s/f(s)$ is a positive increasing function and $\lim_{s\to 0} g(s) = g_0 > \lim_{s\to\infty} g(s) = g_\infty > 0$ are both finite, then the spectrum of problem (6.54) is a finite interval (λ_*, λ^*) in R^+. The situation is simpler in the forced case, i.e., when $f(0) > 0$. We then have

Theorem 6.2 Assume that D is a convex domain of E^N, $f(0) > 0$, and f is increasing for positive arguments. Then problem (6.54) has a positive solution for $\lambda \in (0, \lambda^*)$ and

$$\lambda^* \geq \frac{2}{\rho^2} \sup_{s>0} \frac{s}{f(s)}.$$

Proof We apply Theorem 3.1, using $\underline{u} \equiv 0$ as a subsolution and $\overline{u} = \alpha\psi$ as a supersolution, where

$$\Delta\psi + 1 = 0 \qquad \text{in} \quad D, \tag{6.58}$$

ψ vanishes on ∂D, and α is a positive constant. According to Theorem 3.1, we need

$$\alpha + \lambda f(\alpha\psi) \leq 0 \qquad \text{in} \quad D. \tag{6.59}$$

Since f is increasing, this is satisfied if

$$\alpha + \lambda f(\alpha\psi_M) < 0 \qquad \psi_M = \max_{X \in D} \psi(x).$$

By (6.13) we have $\psi_M \leq \frac{1}{2}\rho^2$, with ρ in the meaning of Theorem 6.3. Thus we need

$$\lambda \leq \alpha / f(\alpha(\rho^2/2)),$$

and making the optimal choice of α we get the sufficient condition

$$\lambda \leq \frac{2}{\rho^2} \sup_{s>0} \frac{s}{f(s)}. \tag{6.60}$$

This implies the statement of Theorem 6.4.

Remark If D is not convex, then one has to take a different bound for ψ_M. In particular, one may use one of the bounds given in Chapter 11, which are valid for nonconvex domains.

6.4.2 Additional Bounds in Special Cases

It was shown by Opial (1961) that $H(s)$ behaves like

$$g(s) = s/f(s) \tag{6.61}$$

in the following sense: If $g(s)$ is strictly increasing (decreasing) on $(0, \infty)$, then the same is true for $H(s)$. Hence the inequality

$$H(u_M) \leq (2\lambda)^{1/2}\rho, \tag{6.62}$$

obtained in the last section, leads to an upper or lower bound, respectively for u_M(for given $\lambda \in (\lambda_*, \lambda^*)$) if $g(s)$ is increasing or decreasing.

For example, if $f(u) = u^p$, $p > 0$, the integral can be evaluated and (6.62) becomes (see also Schaefer and Sperb 1977))

$$u_M{}^{(1-p)/2} N(p) \leq \rho \left(\frac{2\lambda}{p+1} \right)^{1/2}, \tag{6.63}$$

where

$$N(p) = \sqrt{\pi}\, \Gamma\left(\frac{1}{p+1} \right) \left[(p+1)\Gamma\left(\frac{p+3}{2(p+1)} \right) \right]^{-1}.$$

For $0 < p < 1$, (6.63) gives an upper bound for u_M and for $p > 1$, a lower bound, whereas for $p = 1$ we reproduce (6.37). We mention that it follows from a result of Levinson (1962) that problem (6.54) for $f(u) = u^p, p > 1$, has a positive solution for any $\lambda > 0$.

For any $\lambda > 0$, $p > 1$, we thus have by (6.63)

$$u_M \geq \left\{ \frac{N(p)}{\rho(2\lambda/(p+1))^{1/2}} \right\}^{2/(p-1)}.$$

On the other hand, Corollary 5.1 in this special case states that

$$|\nabla u|^2 + \frac{2\lambda}{p+1} u^{p+1} \leq \frac{2\lambda}{p+1} u_M^{p+1}. \tag{6.64}$$

Intergration over D and application of Green's identity yield

$$\left(\lambda + \frac{2\lambda}{p+1} \right) \int_D u^{p+1}\, dx \leq \frac{2\lambda}{p+1} u_M^{p+1} A, \tag{6.65}$$

where A is the "volume" of D, i.e.,

$$u_M \geq \left(\frac{p+3}{2A} \int_D u^{p+1}\, dx \right)^{1/(p+1)}. \tag{6.66}$$

Since (6.66) is valid for any $p > 0$, one could combine (6.63) and (6.66) for $0 < p < 1$ to get

$$\left(\frac{p+3}{2A} \int_D u^{p+1}\, dx \right)^{1/(p+1)} \leq \left(\frac{N(p)}{\rho(2\lambda/(p+1))^{1/2}} \right)^{2/(1-p)}. \tag{6.67}$$

If $H(s)$ is not monotone, (6.62) still contains useful information. For example, if $f(0) > 0$ and $f(s)$ is strictly convex ($f'' > 0$) and increasing for $s > 0$, it was shown by Laetsch (1970) that $H(s)$ has a single maximum for some $s^* > 0$

and $\lim_{s\to\infty} H(s) = 0$. Then for any $\lambda < H^2(s^*)/2\rho^2$ the equation

$$H(s) = (2\lambda)^{1/2}\rho$$

has two solutions $0 < s_1 < s_2 < \infty$. Thus for given $\lambda < H^2(s^*)/2\rho^2$ inequality (6.62) implies that either

$$u_M(\lambda) \geq s_2(\lambda) \tag{6.68}$$

or

$$u_M(\lambda) \leq s_1(\lambda). \tag{6.69}$$

If $f(u) = e^u$ one finds

$$H(s) = e^{-s/2}(s + 2\log(1 + (1 - e^{-s})^{1/2}) \qquad \text{and} \qquad H^2(s^*) = 1.76,$$

i.e., require that $\lambda < 0.88(1/\rho^2)$ in order to apply (6.68) and (6.69). These relations are of course interesting for small values of λ since $s_1(\lambda)$, $s_2(\lambda)$ are far apart then.

6.4.3 Bounds in Terms of the Torsion Function

The maximum principles developed in Chapter 5 can also be applied in order to compare the solution ψ of (6.58) with a solution of (6.54) as was shown among other things in Payne (1980a). One of his comparison results is

Theorem 6.3 Let u be a positive smooth solution of (6.54) in a domain D of E^N. Suppose that

(i) ∂D has nonnegative mean curvature,
(ii) $f(s) \geq 0$, $f'(s) \geq 0$ for $s > 0$.

Set

$$\Phi = \left(\int_0^{u_M} (F(u_M) - F(y))^{-1/2}\,dy\right)^2 - \left(\int_u^{u_M} (F(u_M) - F(y))^{-1/2}\,dy\right)^2.$$

Then the following inequalities hold:

$$\phi \leq 4\lambda\psi \qquad \text{in} \quad D, \tag{6.70}$$

and

$$(F(u_M))^{-1/2} \int_0^{u_M} (F(u_M) - F(y))^{-1/2}\,dy|\nabla u| \leq 2\lambda|\nabla\psi| \qquad \text{on} \quad \partial D. \tag{6.71}$$

Proof In order to establish inequalities (6.70) and (6.71), it suffices to show that ϕ satisfies

$$\Delta\phi + 4\lambda \geq 0 \qquad \text{in} \quad D.$$

The inequalities are then an immediate consequence of the maximum principle. A straightforward calculation gives

$$\Delta\phi + 4\lambda = \left\{2(F(u_M) - F(u))^{1/2} - f(u)\int_u^{u_M} (F(u_M) - F(y))^{-1/2}\,dy\right\}$$
$$\times \left\{\frac{2\lambda(F(u_M) - F(u)) - |\nabla u|^2}{(F(u_M) - F(u))^{3/2}}\right\}.$$

By Corollary 5.1 we have

$$2\lambda(F(u_M) - F(u)) - |\nabla u|^2 \geq 0,$$

so that it remains to show that the first term in braces, call it $B(u)$, is nonnegative. Now

$$B'(u) = -f'(u)\int_u^{u_M} (F(u_M) - F(y))^{-1/2}\,dy \leq 0$$

because of assumption (ii). But $B(u_M) = 0$, and hence $B(u) \geq 0$ for $u \in [0, u_M]$. This proves Theorem 6.1.

Example (a) If $f(u) = u$ and $\lambda = \lambda_1$, i.e., the first positive eigenvalue in the fixed membrane problem, then inequality (6.70) becomes

$$\frac{\pi^2}{4} - \left(\arccos\frac{u}{u_M}\right)^2 \leq 2\lambda_1\psi \quad \text{in} \quad D. \tag{6.72}$$

Evaluated at the point x where $u = u_M$, (6.72) gives

$$\lambda_1\psi_M \geq \lambda_1\psi(x) \geq \frac{\pi^2}{8}, \qquad \psi_M = \max_D \psi. \tag{6.73}$$

On the other hand, (6.71) in our case becomes

$$\pi|\nabla u| < 2\lambda_1 u_M|\nabla\psi| \qquad \text{on} \quad \partial D, \tag{6.74}$$

from which it follows by integration around ∂D and application of Green's identity that

$$E := \frac{\int_D u\,dx}{u_M A} \leq \frac{2}{\pi},$$

as announced in (6.39).

As a further possibility we may combine (6.74) and (6.26) to get (note that $\tau = 2\max_{\partial D}|\nabla\psi|$)

$$\frac{1}{u_M}\max_{\partial D}|\nabla u| \leq \frac{2\lambda_1}{\pi}\rho. \tag{6.75}$$

(b) For $f(u) = u^p$ inequality (6.70) can be written as

$$H(u_M) \leq 2\sqrt{\lambda \psi_M}, \tag{6.76}$$

with the same meaning of $H(u_M)$ as in the previous section. Since $\psi_M \leq \frac{1}{2}\rho^2$ holds, inequality (6.76) is sharper than (6.62).

Various other interesting comparison results can be found in the article of Payne (1980a). In particular, inequalities in problem (6.54) are derived, in which the optimal domain in a disk.

Chapter 7

NONLINEAR ELLIPTIC EQUATIONS IN DIVERGENCE FORM

A general elliptic equation in divergence form can be written as

$$(v(u, u_{,k}, x)u_{,i})_{,i} + f(u, u_{,k}, x) = 0.$$

Although it seems possible to derive a maximum principle for a function $P(u, u_{,k}, x)$ in the spirit of Chapter 5, we shall tackle a more modest task in this chapter and confine our attention to a less general equation of the form

$$(v(q)u_{,i})_{,i} + w(q)f(u) = 0, \qquad q := |\nabla u|^2, \tag{7.1}$$

or in differentiated form

$$v(q)\,\Delta u + 2\frac{dv}{dq}\,u_{,i}u_{,k}u_{,ik} + w(q)f(u) = 0. \tag{7.2}$$

A form more general than that of (7.1) was treated by Payne and Philippin (1980). They considered equations of the form

$$(g(u, q)u_{,i})_{,i} + h(u, q) = 0$$

in place of (7.1). However, a number of practically important problems lead to an equation of the form (7.1).

It is easy to check that the operator

$$L[u] := v(q)\,\Delta u + 2\,\frac{dv}{dq}\,u_{,i}u_{,k}u_{,ik}$$

is uniformly elliptic if for $q \geq 0$ we have

$$v(q) + 2\frac{dv}{dq}q > 0. \tag{7.3}$$

Throughout this chapter it will be tacitly assumed that inequality (7.3) holds, and, additionally, that $w(q) > 0$ for $q \geq 0$.

In the one-dimensional version of (7.1) or (7.2), which we write as

$$v(q)u'' + 2\frac{dv}{dq}qu'' + w(q)f(u) = 0, \qquad q = u'^2, \tag{7.4}$$

the same calculation as the one leading to (5.7) shows that a first integral in (7.4) is

$$\int_0^q \frac{2sv'(s) + v(s)}{w(s)}\,ds + 2\int_0^u f(y)\,dy = \text{const}. \tag{7.5}$$

For the special case $v = w = 1$, we are led back to

$$u'^2 + 2\int_0^u f(y)\,dy = \text{const}.$$

As we have seen in Chapter 5, a particularly simple "P function" satisfying a maximum principle in the special case $v = w \equiv 1$ in (7.1) is

$$P = |\nabla u|^2 + 2\int_0^u f(y)\,dy,$$

or if $f' \leq 0$,

$$P = |\nabla u|^2 + \int_0^u f(y)\,dy.$$

These two special cases in conjunction with (7.5) suggest that we look for functions P of the special form

$$P(\alpha) := \int_0^q \frac{2sv'(s) + v(s)}{w(s)}\,ds + \alpha\int_0^u f(y)\,dy, \tag{7.6}$$

where the constant α is to be determined. Of course one could as in Chapter 5, try a more general form like, e.g.,

$$P = g(|\nabla u|^2) + h(u),$$

and then determine $g(q)$ and $h(u)$ appropriately. For the sake of simplicity we shall confine our attention to the simpler choice made in (7.6) and follow the calculations of Payne and Philippin (1979b).

7.1 MAXIMUM PRINCIPLES

Let u be a sufficiently smooth solution of (7.1) in a finite domain D of E^N, $N \geq 2$, and $P(\alpha)$ be defined by (7.6). Our first goal is to derive an elliptic inequality for P. Since we shall certainly make use of the differential equations (7.1) or (7.2) for u, it is not surprising that we shall try to show that $P(\alpha)$ satisfies an inequality of the form

$$L(P) + W_k P_{,k} \geq 0,$$

with the principal part of the operator L given by

$$\Delta + 2\frac{v'}{v} u_{,i} u_{,k} \frac{\partial^2}{\partial x^i \partial x^k}, \qquad v = v(q), \qquad q = |\nabla u|^2, \qquad v' \equiv \frac{dv}{dq}.$$

We now calculate the derivatives up to the second order of $P(\alpha)$, and we shall suppress the argument α for the time being.

$$P_{,k} = \frac{v + 2qv'}{w} 2u_{,ik} u_{,i} + \alpha f u_{,k}, \tag{7.7}$$

where as before $v = v(q)$, $w(q)$, $f = f(u)$, and primes will denote derivatives with respect to the corresponding arguments. Next we get

$$\begin{aligned} P_{,kj} = &\left\{\frac{3v' + 2qv''}{w} - \frac{w'(v + 2qv')}{w^2}\right\} 4u_{,ik} u_{,i} u_{,lj} u_{,l} \\ &+ \frac{v + 2qv'}{w} 2u_{,ik} u_{,ij} + \frac{v + 2qv'}{w} 2u_{,ikj} u_{,i} \\ &+ \alpha f' u_{,k} u_{,j} + \alpha f u_{,kj}, \end{aligned} \tag{7.8}$$

so in particular

$$\begin{aligned} \Delta P = &\left\{\frac{3v' + 2qv''}{w} - \frac{w'(v + 2qv')}{w^2}\right\} 4u_{,ik} u_{,i} u_{,lk} u_{,l} \\ &+ \frac{v + 2qv'}{w} 2u_{,ik} u_{,ik} + \frac{v + 2qv'}{w} 2u_{,i} \Delta u_{,i} + \alpha f' q + \alpha f \,\Delta u. \end{aligned} \tag{7.9}$$

We first eliminate the term Δu by applying the differential equation (7.2). Then we differentiate (7.2) with respect to x and multiply by $u_{,i}$. Using (7.8), we can eliminate the term $\Delta u_{,i} u_{,i}$ as well. This procedure allows us to derive the relation

$$
\begin{aligned}
&\Delta P + 2\frac{v'}{v}u_{,k}u_{,j}P_{,kj} \\
&\quad = 2\frac{v+2qv'}{w}u_{,ik}u_{,ik} \\
&\qquad + 4u_{,ik}u_{,i}u_{,jk}u_{,j}\left\{\frac{3v'+2qv''}{w} - \frac{w'(v+2qv')}{w^2} - \frac{v'(v+2qv')}{vw}\right\} \\
&\qquad + 8(u_{,ik}u_{,i}u_{,k})^2\left\{\frac{3v'^2}{vw} + \frac{2qv'v''}{vw} - \frac{(2qv'+v)v'w'}{w^2v}\right. \\
&\qquad \left. + \frac{v'^2 - vv''}{wv^2}(v+2qv')\right\} \\
&\qquad + 4u_{,ik}u_{,i}u_{,k}(v+2qv')\left(\frac{v'}{v} - \frac{w'}{w}\right)\frac{f}{v} + \alpha f'q - \frac{\alpha f^2 w}{v} \\
&\qquad + \frac{2\alpha f'v'q^2}{v} - \frac{2f'q(v+2qv')}{v}.
\end{aligned}
\tag{7.10}
$$

We now apply Schwarz's inequality in the form

$$
u_{,ik}u_{,ik}u_{,j}u_{,j} \geq u_{,ik}u_{,i}u_{,jk}u_{,j}. \tag{7.11}
$$

Furthermore we see from (7.7) that the following identities hold

$$
u_{,ik}u_{,i}u_{,k} = -\frac{\alpha vfg}{2(v+2qv')} + A_k P_{,k}, \tag{7.12}
$$

$$
(u_{,ik}u_{,i}u_{,k})^2 = \left\{\frac{\alpha vfg}{2(v+2qv')}\right\}^2 + B_k P_{,k}, \tag{7.13}
$$

$$
u_{,ik}u_{,i}u_{,jk}u_{,j} = \left\{\frac{\alpha vf}{2(v+2qv')}\right\}^2 q + C_k P_{,k}, \tag{7.14}
$$

where the terms A_k, B_k, C_k need not be determined explicitly. Combining (7.10) with (7.11)–(7.14), we are led to

$$
\begin{aligned}
&\Delta P + 2\frac{v'}{v}u_{,k}u_{,j}P_{,kj} + L_k P_{,k} \\
&\quad \geq (\alpha - 2)\left\{(v+2qv')\frac{qf'}{v} + \frac{\alpha}{v^2}\left((v+2qv')wf^2\frac{1}{2} - w'vf^2q\right)\right\}
\end{aligned}
\tag{7.15}
$$

where the term L_k is singular at the point where $\nabla u = 0$. In the special case $\alpha = 2$ we obtain

$$\Delta P + 2\frac{v'}{v}u_{,k}u_{,j}P_{,kj} + L_k P_{,k} \geq 0 \quad \text{in} \quad D. \tag{7.16}$$

Because of our ellipticity assumption (7.3) we may apply the maximum principle to get the counterpart of Lemma 5.1, namely

Lemma 7.1 Let u be a sufficiently smooth solution of (7.2) in a bounded domain D of E^n, $N \geq 2$. Then the function

$$P = \int_0^q \frac{v + 2v's}{w}\,ds + 2\int_0^u f(y)\,dy \tag{7.17}$$

assumes its maximum on ∂D or at a critical point of u.

As we have seen in Chapter 5, the special case in which D is a plane domain allows a somewhat different treatment, since then we may use the identity (5.15) instead of Schwarz's inequality (7.11). The analogous calculations, which are omitted here, now lead to an elliptic equation of the same type, namely,

$$\Delta P + 2\frac{v'}{v}u_{,k}u_{,j}P_{,kj} + \tilde{L}_k P_{,k} = \frac{\alpha - 2}{v}\left\{(v + 2qv')\left(f'q - \frac{wf^2}{v}\right) + \alpha\left[\frac{wf^2}{v}(v + 2qv') - qw'f^2\right]\right\}. \tag{7.18}$$

In the special case $\alpha = 2$ it thus follows that the function P defined by (7.17) must also assume its *minimum* on ∂D or at a critical point ($\tilde{L}_k$ is still singular where $\nabla u = 0$).

Of course we shall try to eliminate one of the possibilities in Lemma 7.1. We shall first give sufficient conditions, ensuring that P as defined by (7.17) assumes its maximum at a critical point. Again the analysis depends on the boundary conditions imposed on u and also on the number of dimensions. The first result is

Theorem 7.1 Let u be a solution of (7.1) in a domain D of E^N satisfying $u = \text{const}$ on ∂D. Suppose that the average curvature H of ∂D is nonnegative. Then the function P defined by (7.17) must assume its maximum at a critical point of u.

Proof Since $u =$ const on ∂D we have

$$\frac{\partial P}{\partial n} = \frac{v + 2qv'}{w} 2 \frac{\partial^2 u}{\partial n^2} \frac{\partial u}{\partial n} + 2f \frac{\partial u}{\partial n}.$$

But on ∂D (7.2) can be rewritten as

$$\frac{\partial^2 u}{\partial n^2}(v + 2qv') + (N-1)Hv \frac{\partial u}{\partial n} + fw = 0,$$

and therefore

$$\frac{\partial P}{\partial n} = -\frac{2(N-1)Hv\left(\dfrac{\partial u}{\partial n}\right)^2}{w} \leq 0$$

since $H \geq 0$ by assumption. The maximum principle and Lemma 7.1 imply that P must assume its maximum where $\nabla u = 0$.

As we have seen in Chapter 5, the analysis becomes more complicated if u is not a constant on the boundary. Although reasoning similar to that given in Chapter 5 could be used in the higher-dimensional case, we shall restrict our attention to the case in which D is a plane domain. In view of the applications considered later on we prove

Theorem 7.2 Let u be a solution of (7.1) in a convex plane domain satisfying the boundary condition

$$v(q) \frac{\partial u}{\partial n} = \gamma = \text{const} > 0. \tag{7.19}$$

Then if $f(s) \leq 0$ for any s, the function P defined by (7.17) assumes its maximum value at a critical point of u.

Proof We assume that P attains its maximum at some point x of ∂D. Then at x we have

$$\frac{\partial P}{\partial s} = 2 \frac{v + 2qv'}{w} \left\{ \frac{\partial u}{\partial n} \frac{\partial}{\partial s}\left(\frac{\partial u}{\partial n}\right) + \frac{\partial u}{\partial s} \frac{\partial^2 u}{\partial s^2} \right\} + 2f \frac{\partial u}{\partial s} = 0. \tag{7.20}$$

Differentiating (7.19) with respect to s, we obtain

$$\frac{\partial}{\partial s}\left(\frac{\partial u}{\partial n}\right) = -\frac{2 \dfrac{\partial u}{\partial n} \dfrac{\partial u}{\partial s} \dfrac{\partial^2 u}{\partial s^2} v'}{v + 2\left(\dfrac{\partial u}{\partial n}\right)^2 v'},$$

and hence (7.20) can be rewritten as

$$\frac{\partial u}{\partial s}\left\{\frac{2\dfrac{\partial^2 u}{\partial s^2}\,v(v+2qv')}{w\left(v+2\left(\dfrac{\partial u}{\partial n}\right)^2 v'\right)}+2f\right\}=0 \qquad \text{at} \quad x. \tag{7.21}$$

We have to investigate the two possible cases

(a) $\partial^2 u/\partial s^2 = -fw(v + 2(\partial u/\partial n)v')/v(v + 2qv')$ at x, or
(b) $\partial u/\partial s = 0$ at x.

In the first case we find

$$\frac{\partial}{\partial s}\left(\frac{\partial u}{\partial n}\right)=\frac{2fwv'\dfrac{\partial u}{\partial n}\dfrac{\partial u}{\partial s}}{v(v+2qv')} \qquad \text{at} \quad x.$$

The term $\partial^2 u/\partial n^2$ can be eliminated, using the fact that the differential equation (7.2) can be written at x in the form

$$\frac{\partial^2 u}{\partial n^2}\left(v+2\left(\frac{\partial u}{\partial n}\right)^2 v'\right)=-wf-\frac{\partial^2 u}{\partial s^2}\left(v+\ \frac{\partial u}{\partial s}v'\right)-4v'\frac{\partial u}{\partial s}\frac{\partial u}{\partial n}\frac{\partial}{\partial s}\left(\frac{\partial u}{\partial n}\right)$$
$$-vk\frac{\partial u}{\partial n}+2kv'\frac{\partial u}{\partial n}\left(\frac{\partial u}{\partial s}\right)^2,$$

where k is the curvature of ∂D at x. Now we can write

$$\frac{\partial P}{\partial n}=2\frac{v+2qv'}{w}\frac{\partial^2 u}{\partial n^2}\frac{\partial u}{\partial n}+\frac{\partial}{\partial s}\left(\frac{\partial u}{\partial n}\right)-k\frac{\partial u}{\partial s}+2f\frac{\partial u}{\partial n},$$

and after some algebra it turns out that

$$\frac{\partial P}{\partial n}=\frac{2}{wv}\frac{v+2qv'}{v+2\left(\dfrac{\partial u}{\partial n}\right)^2 v'}(fw\gamma-kv^2q)\le 0 \tag{7.22}$$

since by assumption $f \le 0$ and $k \ge 0$.

In case (b) we find, since $(\partial u/\partial s) = \partial/\partial s(\partial u/\partial n) = 0$ on ∂D

$$\frac{\partial P}{\partial n}=-\frac{2\gamma}{w}\left(k\frac{\partial u}{\partial n}+\frac{\partial^2 u}{\partial s^2}\right).$$

It is clear that if $\partial u/\partial s = 0$ on ∂D, then by (7.17) and (7.19) we have on ∂D

$$P = \text{const} + 2\int_0^u f(y)\,dy,$$

and therefore

$$\frac{dP}{du} = 2f(u) \leq 0,$$

i.e., the maximum of P on ∂D must occur where u itself is a minimum on ∂D, i.e.,

$$\frac{\partial^2 u}{\partial s^2} \geq 0 \qquad \text{at} \quad x,$$

and hence

$$\frac{\partial P}{\partial n} \leq 0 \qquad \text{at} \quad x,$$

also in case (b). This establishes Theorem 7.2.

Remarks (a) The condition $f(u) \leq 0$ is clearly compatible with the condition necessary for the existence of solutions to (7.1) satisfying (7.19), since by the divergence theorem we have

$$\int_D (vu_{,i})_{,i}\, dx + \int_D wf\, dx = \oint_{\partial D} v \frac{\partial u}{\partial n}\, ds + \int_D wf\, dx = \gamma L + \int_D wf\, dx = 0,$$

where L is the length of ∂D. Since $\gamma > 0$ and $w > 0$ by assumption, we see that $f(u) \geq 0$ is impossible. In many applications f does not change sign, and therefore $f(u) \leq 0$, $f(u) \not\equiv 0$ is a natural requirement.

(b) A more thorough analysis shows that in Theorems 7.1 and 7.2 the assumption $H \geq 0$ $(k \geq 0)$ can be dropped as long as $H(k)$ stays bounded by an appropriate choice of $\alpha > 2$.

So far we have used only the special value $\alpha = 2$. As we shall see, a different choice of the value of α (for $N > 2$) leads to a different result. But first we need an extension of Lemma 5.4 to solutions of (7.1), namely,

Lemma 7.2 Let u be a solution of (7.1). Then the inequality

$$u_{,ij}u_{,ij} \geq \frac{1}{N}\frac{f^2w^2}{v} - \frac{4v'}{v}\left(1 + \frac{qv'}{v}\right)u_{,i}u_{,ij}u_{,k}u_{,kj} \tag{7.23}$$

holds in $D \subset E^N$.

Proof Consider

$$Q_{ij} := \left(u_{,ij} + 2\frac{v'}{v}u_{,i}u_{,k}u_{,kj}\right) - \frac{1}{N}\left(\Delta u + 2\frac{v'}{v}u_{,kl}u_{,k}u_{,l}\right)\delta_{ij}.$$

Now one simply uses the fact that

$$Q_{ij}Q_{ij} \geq 0,$$

observing the differential equation (7.1). This gives

$$\left(u_{,ij} + 2\frac{v'}{v}\,u_{,i}u_{,k}u_{,kj}\right)\left(u_{,ij} + 2\frac{v'}{v}\,u_{,i}u_{,k}u_{,kj}\right) \geq \frac{1}{N}\frac{f^2w^2}{v^2},$$

and when solved for $u_{,ij}u_{,ij}$, we obtain (7.23).

We are now able to prove

Theorem 7.3 Let u be a solution of (7.1) in some finite domain D of E^N. If

$$\frac{2}{N}\,w'f^2 - (v + 2qv')f' \geq 0,$$

then the function

$$P := \int_0^q \frac{v(\xi) + 2\xi v'(\xi)}{w(\xi)}\,d\xi + \frac{2}{N}\int_0^q f(y)\,dy \tag{7.24}$$

takes its maximum on ∂D.

Proof We go back to equation (7.10) and use (7.23) in conjuction with the identities (7.12)–(7.14). After some routine calculations, one now finds instead of (7.15)

$$\Delta P + 2\frac{v'}{v}\,u_{,k}u_{,j}P_{,kj} + \hat{L}_kP_{,k}$$

$$\geq \alpha(2-\alpha)\frac{w'f^2q}{v} + \frac{v + 2qv'}{v}\left\{\frac{f^2w}{v}\left(\frac{2}{N} - \alpha\right) + f'q(\alpha - 2)\right\}, \tag{7.25}$$

where $\hat{L}_k$ is bounded in D. The choice of $\alpha = 2/N$ along with the assumption in Theorem 7.3 shows that P as defined by (7.24) satisfies an elliptic differential inequality so that the maximum principle can be applied.

Remarks (a) It is important to note that no convexity assumption had to be made, and no boundary conditions were imposed on u.

(b) For $\alpha = 0$ we see from (7.25) that $|\nabla u|$ is a maximum on ∂D if

$$\frac{1}{N}\,f^2(u)w(q) \geq f'(u)qv(q),$$

which is satisfied, e.g., if $f' \leq 0$.

7.2 APPLICATIONS

7.2.1 Torsional Creep

The problem of torsional creep (see, e.g., the papers of Anderson and Arthurs, (1972) and Langenbach, (1964)) leads to the equation

$$(v(q)u_{,i})_{,i} + 2 = 0 \qquad \text{in} \quad D \subset E^2, \qquad q := |\nabla u|, \tag{7.26}$$

with u vanishing on ∂D, i.e., the special case of (7.1) in which $w(q) \equiv 2$. Note that for $v(q) \equiv 1$ (7.26) reduces to the equation of the classical Saint Venant torsion problem. Again $v(q)$ is assumed to satisfy the ellipticity condition

$$v(q) + 2qv'(q) > 0, \qquad v(q) > 0.$$

We first derive a bound for the maximum shear strain energy defined as $Q = \max_{\partial D} (\sqrt{q}\, v(q))$, based on Theorem 7.3. In our case, it states that

$$P := \int_0^q (v(\xi) + 2v'(\xi)\xi)\, d\xi + 2u \tag{7.27}$$

attains its maximum on ∂D. By the maximum principle we have at a point x of ∂D, where P assumes its maximum,

$$\frac{\partial P}{\partial n} > 0,$$

i.e.,

$$(v(q) + 2v'(q)q)\frac{\partial q}{\partial n} + 2\frac{\partial u}{\partial n} > 0.$$

But

$$q = \left(\frac{\partial u}{\partial n}\right)^2 \qquad \text{on} \quad \partial D,$$

so that

$$\frac{\partial q}{\partial n} = 2\frac{\partial^2 u}{\partial n^2}\frac{\partial u}{\partial n}.$$

In addition, (7.27) on ∂D can be written as

$$(v + 2qv')\frac{\partial^2 u}{\partial n^2} = -2 - vk(M)\frac{\partial u}{\partial n}.$$

Combining our last four relations, we thus get (see Payne and Philippin (1977a))

$$Q = \max_{\partial D} (\sqrt{q}\, v(q)) < \frac{1}{k(M)} \leq \frac{1}{k_0}, \tag{7.28}$$

where $k(x) \geq k_0 > 0$ is supposed to hold, i.e., D is strictly convex. Using a purely geometric inequality, Bandle (1979) derived (7.28) as well as the lower bound

$$Q \geq \frac{1}{k_m}, \qquad k_m = \max_{\partial D} k. \tag{7.29}$$

On the other hand, we find

$$\oint_{\partial D} v(q) \frac{\partial u}{\partial n} \, ds + 2A = -\oint_{\partial D} v(q) \sqrt{q} \, ds + 2A = 0$$

if we just integrate (7.26) over D, using the divergence theorem. Hence

$$Q \geq \frac{2A}{L}, \qquad A = \text{area of } D, \qquad L = \text{length of } \partial D. \tag{7.30}$$

But because of the fact that

$$k_m L \geq \oint_{\partial D} k \, ds = 2\pi$$

and because of the classical isoperimetric inequality $L^2 \geq 4\pi A$, it follows that (7.30) is sharper than the lower bound given in (7.29).

In Theorem 7.3 no convexity assumption had to be made. Hence the calculations leading to (7.28) also contain information about the location of the point x where $\sqrt{q}\, v(q)$ assumes its maximum Q. A combination of (7.28) and (7.30) shows that at the point x we must have

$$k(x) \leq L/2A. \tag{7.31}$$

Note that (7.31) is independent of $v(q)$.

So far we have only applied Theorem 7.3. Further bounds in problem (7.26) can be obtained as applications of Theorem 7.1. It tells us that if we chose

$$P := \int_0^q (v(y) + 2yv'(y)) \, dy + 4u$$

instead of (7.27), then for convex domains D, P assumes its maximum where $\nabla u = 0$, which may be expressed in the inequality

$$\int_0^q (v(y) + 2yv'(y)) \, dy \leq 4(u_M - u), \qquad u_M = \max_D u. \tag{7.32}$$

Following Payne and Philippin (1977a), we assume that $v(y)$ satisfies

$$v(y) + 2yv'(y) \geq cy^m, \qquad 0 \leq y \leq q_M, \qquad q_M = \max_{\partial D} q, \tag{7.33}$$

where c and m are positive constants. The procedure that led to (6.13) can be repeated here. Inequality (7.32) together with (7.33) yields

$$q^{2m+2} \leq 4\,\frac{m+1}{c}\,(u_M - u), \tag{7.34}$$

so that in the notation of Section 6.1, we find after an easy integration that

$$\rho \geq \left(\frac{c}{4(m+1)}\right)^{1/(2m+2)} \frac{2m+2}{2m+1}\,(u_M)^{(2m+1)/(2m+2)} \tag{7.35}$$

and therefore

$$q_M \leq \left(\frac{2\rho}{c}\,(2m+1)\right)^{1/(2m+1)} \tag{7.36}$$

Bounds similar to (7.36) can be established under different assumptions on $v(q)$ (see Payne and Philippin, (1977a)).

As a last application we take the case in which $v(y) = y^m$. The "torsional stiffness" of D is defined as

$$S_v := \int_D |\nabla u|^2 v(|\nabla u|^2)\,dx = 2\int_D u\,dx, \tag{7.37}$$

which for $v(y) = y^m$ becomes

$$S_m := \int_D |\nabla u|^{2m+2}\,dx. \tag{7.38}$$

Then integration of inequality (7.32) immediately gives

$$S_m \leq \frac{4(m+1)}{3+4m}\,u_M A. \tag{7.39}$$

Finally, (7.35) and (7.39) lead to ($c = 2m+1$ here)

$$S_m \leq \frac{2(2m+1)(m+1)}{3+4m}\,2^{1/(2m+1)}\rho^{(2m+2)/(2m+1)}A. \tag{7.40}$$

Additional inequalities for S_m can also be found in the papers of Payne and Philippin (1977a,b).

7.2.2 Surface of Constant Mean Curvature H

Let D be a plane domain and Σ a surface given in explicit form by

$$\Sigma:\quad x^3 = u(x^1, x^2), \qquad (x^1, x^2) \in D \subset E^2.$$

Suppose now that Σ has constant mean curvature H and furthermore contains ∂D. This means that we are seeking a solution of

$$\left(\frac{u_{,i}}{\sqrt{1+|\nabla u|^2}}\right)_{,i} = 2H = \text{const} \qquad \text{in} \quad D, \tag{7.41}$$

with u vanishing on ∂D. It was shown by Serrin (1969) that equation (7.41) with $u = \gamma(s)$, $\gamma(s)$ an arbitrary C^2 function on ∂D, has a unique solution if the following condition holds:

$$k(s) \geq 2H \qquad \text{on} \quad \partial D, \tag{7.42}$$

where $k(s)$ is the curvature of ∂D. If (7.42) does not hold—even at a single point—then there are boundary data $\gamma(s) \subset C^2$ such that Eq. (7.41) with $u = \gamma(s)$ on ∂D is not solvable. In the special case that $\gamma \equiv 0$ it still follows from a result of Bernstein (1927) that (7.41) can have no solution if D contains a circle of radius H^{-1}. We now apply Theorem 7.3 with $v(\xi) = (1 + \xi)^{-1/2}$, $w(\xi) \equiv 1$, and $f(\xi) \equiv -2H$. Thus P as defined in (7.24) becomes

$$P := 2(1 - (1 + q)^{-1/2}) + Hu \tag{7.43}$$

According to Theorem 7.3, P assumes its maximum at some point x on ∂D. Again, we make use of the fact that $\partial P/\partial n > 0$ at x. Analogously as before we now find

$$k_0 \leq k(P) \leq \frac{H(1 + q_M)^{1/2}}{q_M}, \qquad q_M = \max_{\partial D} |\nabla u|^2. \tag{7.44}$$

On the other hand, since P as defined by (7.43) assumes its maximum on ∂D, we also have

$$Hu_M \leq 1 - (1 + q_M)^{-1/2}. \tag{7.45}$$

A combination of (7.44) and (7.45) gives

$$Hu_M \leq 1 - \frac{(k_0^2 - H^2)^{1/2}}{k_0}. \tag{7.46}$$

It is not hard to check that the equality signs in (7.44)–(7.46) hold if ∂D is a circle.

Let us see now what information Theorem 7.1 contains about problem (7.41). According to Theorem 7.1, the slightly modified function

$$P := 2(1 - (1 + q)^{-1/2} + 2Hu) \tag{7.47}$$

attains its maximum at a critical point of u, so that

$$(1 + q)^{-1/2} \geq 1 - 2H(u_M - u) \qquad \text{in} \quad D. \tag{7.48}$$

The right-hand side in (7.48) is nonnegative in D, provided that

$$2Hu_M \leq 1, \tag{7.49}$$

which by (7.46) holds if

$$k_0 \geq \frac{2}{\sqrt{3}} H. \tag{7.50}$$

The condition (7.50) is therefore weaker than Serrin's existence criterion (7.41). If (7.50) holds, we may square both sides in (7.48), so that we are led to

$$q^{1/2} = |\nabla u| \leq [(1 - 2H(u_M - u))^{-2} - 1]^{1/2}. \tag{7.51}$$

Again we perform an integration of (7.51) from the point at which $u = u_M$ to the nearest boundary point and find

$$\rho \geq \frac{1}{2H}(1 - (1 - 2Hu_M)^2)^{1/2}, \tag{7.52}$$

i.e.,

$$2Hu_M \leq 1 - (1 - 4H^2\rho^2)^{1/2}, \tag{7.53}$$

provided $2H\rho < 1$.

We can use (7.48) also as follows: Since

$$\frac{1}{\sqrt{1+q}} = \sqrt{1+q} - \frac{q}{\sqrt{1+q}},$$

we obtain from (7.48)

$$2Hu \leq \sqrt{1+q} - \frac{q}{\sqrt{1+q}}. \tag{7.54}$$

Using the fact that $q/\sqrt{1+q} = (\nabla u, \mathbf{v})$, with $\mathbf{v} = \nabla u/\sqrt{1 + |\nabla u|^2}$ and the divergence theorem in conjunction with (7.40), we find from (7.54) after an integration over D that

$$|\Sigma| \geq 4HV, \tag{7.55}$$

where $|\Sigma| := \int_D \sqrt{1 + |\nabla u|^2}\, dx$ is the surface area of Σ and V the volume between D and Σ. Note that (7.55) holds for domains D such that $2\Lambda u_M \leq 1$. On the other hand, the same integration could also be used for P as defined by (7.43). Then

$$Hu \leq \frac{1}{\sqrt{1+q}} = \sqrt{1+q} - \frac{q}{\sqrt{1+q}} \tag{7.56}$$

is the inequality corresponding to (7.54), and the integration now yields

$$|\Sigma| \geq 3HV. \tag{7.57}$$

Note that (7.57) and (7.55) hold in different classes of domains. Inequality (7.55) is optimal in the sense that for a rectangular strip of width H^{-1} we have $4HV|\Sigma|^{-1} \to 1$ as its length tends to infinity. In contrast, (7.57) is optimal in the sense that the equality sign in (7.57) holds if D is a disk of radius H^{-1} (see Theorem 5 of Payne and Philippin (1979b)).

Various other bounds could be derived as well. For example, from (7.43) we know that

$$Hu \leq \frac{1}{\sqrt{1+q}} - \frac{1}{\sqrt{1+q_M}}; \tag{7.58}$$

hence after an integration over D and some manipulation we get

$$1 + q_M \geq \left(\frac{A}{|\Sigma| - 3HV}\right)^2, \qquad q_M = \max_{\partial D} |\nabla u|^2. \tag{7.59}$$

The equality sign in (7.59) again holds if D is a disk. Finally, one could combine (7.59) and (7.51), evaluated on ∂D, to get the nonisoperimetric result

$$AHu_M > A + 3HV - |\Sigma|. \tag{7.60}$$

For further bounds in the problem of a surface of constant mean curvature refer to Chapter 11.

7.2.3 Capillary Surface

This problem has been mentioned in the introduction. Again, D is a plane domain, and the fluid surface is given by $x^3 = u(x^1, x^2)$ where u satisfies the equation

$$\left(\frac{u_{,i}}{\sqrt{1+|\nabla u|^2}}\right)_{,i} = cu \tag{7.61}$$

and c is a positive constant whose physical interpretation is explained in Section 1.2.3. Two different boundary conditions are of special interest in this context:

(a) *The meniscus problem* The capillary tube is so short that the fluid rises to the top along the rim of the tube. The boundary condition for u then is

$$u = a = \text{const} > 0 \qquad \text{on} \quad \partial D. \tag{7.62}$$

(b) *The long capillary tube* The fluid does not reach the top anywhere so that the boundary condition becomes

$$(1+q)^{-1/2}\frac{\partial u}{\partial n} = \cos\gamma \qquad \text{on} \quad \partial D, \tag{7.63}$$

where γ is the wetting angle and $0 < \gamma < \frac{1}{2}\pi$.

Serrin (1969) has shown that problem (7.61), (7.62), with the constant a replaced by a function $\phi(s)$, has a unique solution provided that

$$|\phi| \le \frac{k(s)}{c} \qquad \text{on} \quad \partial D. \tag{7.64}$$

For existence results in the long capillary tube problem the reader is referred, e.g., to the papers of Simon and Spruck (1976), Emmer (1973), Ural'ceva (1973), and Concus and Finn (1969; 1974a,b). For both cases (a) and (b), we know by Theorem 7.3 that

$$P := 2\left(1 - \frac{1}{\sqrt{1+q}}\right) - \frac{1}{2}cu^2 \tag{7.65}$$

assumes its maximum on ∂D. In case (a) we get the relation

$$\frac{c}{4}(a^2 - u^2) \le \frac{1}{\sqrt{1+q}} - \frac{1}{\sqrt{1+q_M}} \qquad \text{in} \quad D. \tag{7.66}$$

Evaluating (7.66), .e.g, at the point of the minimum rise of the fluid, yields

$$\frac{c}{4}(a^2 - u_m^2) \le 1 - \frac{1}{\sqrt{1+q_M}}, \qquad u_m = \min_D u. \tag{7.67}$$

As before we can take advantage of the fact that $\partial P/\partial n > 0$ at the point of ∂D where P is a maximum. The usual calculation now shows that

$$q_M \le \left(\frac{4k_0^2}{c^2a^2} - 1\right)^{-1/2}, \qquad k(s) \ge k_0 > 0, \tag{7.68}$$

which can be combined with (7.67) to give

$$u_m^2 \ge 4\left\{\sqrt{\frac{1}{c^2} - \frac{a^2}{4k_0^2}} + \frac{a^2}{4} - \frac{1}{c}\right\}. \tag{7.69}$$

We also may apply Theorem 7.1 in case (a). Stated as an inequality, Theorem 7.1 tells us that

$$\frac{1}{\sqrt{1+q}} \ge 1 - \frac{c}{2}(u^2 - u_m^2) \qquad \text{in} \quad D. \tag{7.70}$$

Note that the right-hand side in (7.70) is positive provided that $\frac{1}{2}c(a^2 - u_m^2) \leq 1$ which will be true if

$$ac \leq \sqrt{3}k_0 \tag{7.71}$$

because of (7.69). This condition, however, is weaker than Serrin's existence criterion

$$ac \leq k_0. \tag{7.72}$$

If (7.71) is satisfied, we may square inequality (7.70) and obtain

$$q = |\nabla u|^2 \leq [1 - \tfrac{1}{2}c(u^2 - u_m^2)]^{-1} - 1 \leq (1 - \tfrac{1}{2}cu^2)^{-2} - 1. \tag{7.73}$$

As before we now integrate from a point where $\nabla u = 0$ and $u = u_0$ to the nearest point of ∂D. This gives

$$\rho\sqrt{2c} \geq \log \frac{a^2}{u_0^2} - \frac{c}{2}(a^2 - u_0^2), \tag{7.74}$$

which can be used to derive a lower bound for u_0 in terms of ρ (the radius of of the largest inscribed circle).

Finally we turn our attention to the long capillary tube problem (b). Again we use the fact that P as defined by (7.65) assumes its maximum on ∂D. From the proof of Theorem 7.2 we know, furthermore, that $u = \min_{\partial D} u$ at the point where P assumes its maximum; hence $\partial u/\partial s = 0$ there, and therefore $q = (\partial u/\partial n)^2$. By (7.63) we see that where $P = P_{\max}$, we have

$$\frac{1}{\sqrt{1+q}} = \sin\gamma. \tag{7.75}$$

But then P takes the form

$$P = 2\left(1 - \sin\gamma - \frac{k_0^2}{c}\cos^2\gamma\right), \tag{7.76}$$

so that by Theorem 7.3 we have

$$\frac{1}{\sqrt{1+q}} + \frac{c}{4}u^2 \geq \sin\gamma + \frac{k_0^2}{c}\cos^2\gamma \qquad \text{in} \quad D, \tag{7.77}$$

and hence

$$\frac{c}{4}u_m^2 \geq \sin\gamma - 1 + \frac{k_0^2}{c}\cos^2\gamma, \tag{7.78}$$

which is of interest if the right-hand side in (7.78) is positive. Of course various other bounds could be derived along the same lines by combinations of Theorems 7.2 and 7.3.

7.2.4 Concluding Remarks

In Payne (1980b) a number of results were proven that are, in spirit, closely related to the results of this section, and in particular to the results given in Section 6.4.3. Payne considered the problem

$$(g(|\nabla u|^2)u_{,i} + 2k = 0 \quad \text{in} \quad D \subset E^2, \qquad u = 0 \quad \text{on} \quad \partial D. \tag{7.79}$$

Here k is a positive constant and $g(s)$ satisfies for $s \geq 0$

$$g(s) > 0 \qquad \text{and} \qquad g(s) + 2sg'(s) > 0.$$

He then proved a number of comparison results, comparing the solution of (7.79) with the torsion function. For example, in the problem of a surface of a constant mean curvature he showed that in the notation of Section 7.2.1 one has

$$V/A \leq \pi/8H,$$

with equality if D degenerates to a strip.

Chapter 8

THE *P*-FUNCTION FOR SOLUTIONS OF $Lu + f(u) = 0$

The main idea of this chapter is to introduce a geometric interpretation of the results of Chapter 5 in order to extend them to the case in which the Laplacian is replaced by a uniformly elliptic operator. Employing a geometric interpretation, the form of the corresponding function P can be found without difficulty. The principal part of this chapter is given in Sperb (1979a).

8.1 ELLIPTIC OPERATOR AND LAPLACE–BELTRAMI OPERATOR

Many problems occurring in physics and other sciences have differential operators that are derived from a variational principle. It is therefore natural to assume that for any C^1 function $v(x^k)$ the quadratic form

$$Q(x^k) := a^{ij}(x^k)v_{,i}v_{,j}$$

is invariant with respect to transformations of the coordinates x^k. Furthermore, we shall assume that for some $\mu_0 > 0$ we have

$$a^{ij}\xi_i\xi_j \geq \mu_0\xi_i\xi_i,$$

so that the operator

$$Lu := a^{ij}(x^k)\frac{\partial^2 u}{\partial x^i\,\partial x^j} + b^i(x^k)\frac{\partial u}{\partial x^i} \equiv a^{ij}u_{,ij} + b^i u_{,i}$$

is uniformly elliptic. In addition, $a^{ij} = a^{ji}$ will be required throughout, and hence a^{ij} is a positive definite symmetric tensor. In the place of Eq. (5.9) we are now concerned with the differential equation

$$Lu + f(u) = 0 \qquad \text{in} \quad D \subset E^N. \tag{8.1}$$

A natural way to interpret equation (8.1) geometrically is as follows. We set

$$g_{ij} = (a^{ij})^{-1}$$

and consider g_{ij} as the covariant metric tensor of a Riemannian space R^N in which the line element is therefore given by

$$ds^2 = g_{ij}\,dx^i\,dx^j = (a^{ij})^{-1}\,dx^i\,dx^j.$$

Let $\tilde{\Delta}$ be the Laplace–Beltrami operator in R^N. Then we have

$$\tilde{\Delta}u = \frac{1}{\sqrt{g}}\frac{\partial}{\partial x^i}\left(\sqrt{g}\,g^{ij}\frac{\partial u}{\partial x^j}\right), \qquad g = \det(g_{ij}), \qquad g^{ij} = (g_{ij})^{-1},$$

or in differentiated form

$$\tilde{\Delta}u = g^{ij}\frac{\partial^2 u}{\partial x^i \partial x^j} + \frac{1}{\sqrt{g}}\frac{\partial}{\partial x^i}(\sqrt{g}\,g^{ij})\frac{\partial u}{\partial x^j},$$

which in terms of our original elliptic operator becomes

$$\tilde{\Delta}u = Lu + t^j u_{,j}, \tag{8.2}$$

with

$$t^j := \sqrt{a}\,\frac{\partial}{\partial x^i}\left(\frac{1}{\sqrt{a}}\,a^{ij}\right), \qquad a = \det(a^{ij}). \tag{8.3}$$

Of course the domain $D \subset E^N$ can also be interpreted as a domain in R^N so that it becomes evident now that Eq. (8.1) is equivalent to

$$\tilde{\Delta}u + \frac{\partial u}{\partial r} + f(u) = 0 \qquad \text{in} \quad D, \tag{8.4}$$

where

$$\frac{\partial u}{\partial r} := r^i u_{,i}, \qquad r^i = b^i - t^i = b^i - \sqrt{a}\,\frac{\partial}{\partial x^j}\left(\frac{1}{\sqrt{a}}\,a^{ij}\right). \tag{8.5}$$

Finally, let us look at the usual boundary conditions. The conormal derivative $\partial u/\partial N$ (N is not the dimension here) of u with respect to the operator L is defined by

$$\frac{\partial u}{\partial N} = a^{ij} n_j \frac{\partial u}{\partial x^i}, \qquad n_j = \text{Euclidean unit normal on } \partial D. \tag{8.6}$$

On the other hand, the unit normal v on ∂D in the Riemannian sense is given by

$$v^i = \frac{g^{ij} n_j}{(g^{ij} n_i n_j)^{1/2}} = \frac{a^{ij} n_j}{(a^{ij} n_i n_j)^{1/2}},$$

and the normal derivative in R^N by

$$\frac{\partial u}{\partial v} = v^i u_{,i}. \tag{8.7}$$

Hence, a comparison shows that the following two problems are equivalent:

$$\begin{aligned} Lu + f(u) &= 0 \qquad \text{in} \quad D, \\ \frac{\partial u}{\partial N} + \gamma\sigma(u) &= 0 \qquad \text{on} \quad \partial D, \end{aligned} \tag{8.8}$$

and

$$\begin{aligned} \tilde{\Delta} u + \frac{\partial u}{\partial r} + f(u) &= 0 \qquad \text{in} \quad D, \\ \frac{\partial u}{\partial v} + \sigma(u) &= 0 \qquad \text{on} \quad \partial D, \end{aligned} \tag{8.9}$$

with $\gamma = (a^{ij} n_i n_j)^{1/2}$ and $\partial u/\partial r$ given by (8.5).

In particular the Dirichlet and Neumann problems (with $\partial u/\partial N$ in the place of $\partial u/\partial n$) correspond to each other. In the next section we shall extend the results of Chapter 5 based on the equivalence of (8.8) and (8.9). The development will roughly parallel that of Chapter 5.

8.2 MAXIMUM PRINCIPLES

Throughout this chapter a comma followed by a subscript will now denote a covariant derivative, whereas superscripts will be used for contravariant derivatives. Thus, for instance

$$|\tilde{\nabla} u|^2 := u^{,i} u_{,i} = g^{ij} u_{,j} u_{,i}$$

is the square of the gradient in R^N, and the scalar product is given by

$$\tilde{\nabla} u \cdot \tilde{\nabla} v = u^{,i} v_{,i} = u_{,j} v^{,j} = g^{ij} u_{,i} v_{,j}.$$

Let u be a $C^3(D) \cap C^2(\bar{D})$ solution of (8.1) or, equivalently, of (8.4). As in Chapter 5 we now set

$$P = |\tilde{\nabla} u|^2 g(u) + h(u)$$

and try to choose $g(u)$ and $h(u)$ such that P satisfies a maximum principle. Again the two-dimensional case allows a somewhat different treatment.

8.2.1 Elliptic Inequalities for P

We calculate the derivatives of P. First, we have

$$P_{,i} = 2u^{,j}{}_{,i} u_{,j} g + |\tilde{\nabla} u|^2 g' u_{,i} + h' u_{,i}, \tag{8.10}$$

and

$$\begin{aligned} \tilde{\Delta} P = P_{,i}{}^{,i} = {} & 2u^{,j}{}_{,i} u_{,j}{}^{,i} g + 2u^{,j}{}_{,i}{}^{,i} u_{,j} g + 4u^{,j}{}_{,i} u_{,j} u^{,i} g' \\ & + |\tilde{\nabla} u|^4 g'' + |\tilde{\nabla} u|^2 (g' \tilde{\Delta} u + h'') + h' \tilde{\Delta} u. \end{aligned} \tag{8.11}$$

In order to eliminate the third-derivative term in (8.11), we make use of the Ricci identity (4.27). It states that

$$u_{,klm} = u_{,lmk} + R^s{}_{lkm} u_{,s},$$

where $R^s{}_{lkm}$ is the mixed Riemann–Christoffel tensor. We can now raise the indices and relabel them to get

$$u_{,ji}{}^{,i} = (\tilde{\Delta} u)_{,j} + g^{ip} R^s{}_{ijp} u_{,s}.$$

Using the identities (4.28), (4.30), and (4.34), we can write

$$\begin{aligned} u^{,j}{}_{,i}{}^{,i} u_{,j} = u_{,ji}{}^{,i} u^{,j} &= ((\tilde{\Delta} u)_{,j} - R^s_j u_{,s}) u^{,j} \\ &= (\tilde{\Delta} u)_{,j} u^{,j} - R_{sj} u^{,s} u^{,j}, \end{aligned} \tag{8.12}$$

where R_{sj} is the Ricci tensor of R^N. By the differential equation (8.4) we have

$$(\tilde{\Delta} u)_{,j} u^{,j} = -f' |\tilde{\nabla} u|^2 - r^i_{,j} u_{,i} u^{,j} - r^i u_{,ij} u^{,j},$$

and hence

$$\begin{aligned} \tilde{\Delta} P + \frac{\partial P}{\partial r} = {} & 2u_{,ij} u^{,ij} g + |\tilde{\nabla} u|^2 (h'' - fg' - 2f'g) + 4u_{,ij} u^{,i} u^{,j} \\ & + |\tilde{\nabla} u|^4 g'' - h'f - 2g T_{sj} u^{,s} u^{,j}, \end{aligned} \tag{8.13}$$

where we have set

$$T_{sj} := R_{sj} + g_{kj} r^k{}_{,s}. \tag{8.14}$$

By (8.11) we can also write

$$4g'u^{,j}{}_{,i}u^{,i}u_{,j} = 2\frac{g'}{g}(P_{,i} - |\tilde{\nabla}u|^2 g'u_{,i} - h'u_{,i})u^{,i}$$

$$= 2\frac{g'}{g}(P_{,i}u^{,i} - g'|\tilde{\nabla}u|^4 - h'|\tilde{\nabla}u|^2), \tag{8.15}$$

which allows to put (8.13) into the form

$$\tilde{\Delta}P + W^iP_{,i} = 2u^{,ij}u_{,ij}g + |\tilde{\nabla}u|^4\left(g'' - 2\frac{g'^2}{g}\right) + |\tilde{\nabla}u|^2\left(h'' - f'g\right.$$

$$\left. - 2f'g - 2\frac{g'h'}{g}\right) - h'f - 2gT_{sj}u^{,s}u^{,j}, \tag{8.16}$$

with $W^i = r^i - 2(g'/g)u^{,i}$. This is the analog of Eq. (5.14). Note that up to this point the dimension N does not play any role. Let us first look at the case $N = 2$. In order to eliminate the term $2u^{,ij}u_{,ij}g$ we prove an analog of the identity (5.15).

Lemma 8.1 For any sufficiently smooth function v in $D \subset R^2$ we have

$$|\tilde{\nabla}v|^2v^{,ij}v_{,ij} = |\tilde{\nabla}v|^2(\tilde{\Delta}v)^2 + 2v^{,i}v_{,j}v_{,ik}v^{,jk} - 2\,\tilde{\Delta}v v_{,i}v_{,j}v^{,ij}. \tag{8.17}$$

Proof We rewrite (8.17) as

$$v^{,i}{}_{,j}v_{,i}{}^{,j} = (\tilde{\Delta}v)^2 + 2\,\frac{v^{,i}v_{,j}v_{,ik}v^{,jk} - \tilde{\Delta}v\, v^{,i}{}_{,j}v_{,i}v^{,j}}{|\tilde{\nabla}v|^2}.$$

Now, $\tilde{\Delta}v = v^{,i}{}_{,i} = v_{,i}{}^{,i}$, so that it is easy to check that the terms with $i = j$ are identical on both sides. For the terms in which $i \neq j$ one just has to use that $(i,j) = (1,2)$ or $(2,1)$. This shows that the identity (8.17) holds. By means of the identity (8.17) as well as (8.11) and (8.16) we find the analog of (5.17) to be

$$\tilde{\Delta}P + \frac{L^iP_{,i}}{|\tilde{\nabla}u|^2} = g(\log g)''|\tilde{\nabla}u|^4 + ((h' - 2fg)' - fg')|\tilde{\nabla}u|^2$$

$$- 2gT_{sj}u^{,s}u^{,j} + \frac{1}{g}(h' - fg)(h' - 2fg), \tag{8.18}$$

where $L^i = -(1/g)(P^{,i} - 2u^{,i}(h' - fg) - r^i|\tilde{\nabla}u|^2)$. We can also take advantage of the fact that the Ricci tensor degenerates in a two-dimensional Riemannian space, so that (see Chapter 4)

$$R_{sj}u^{,s}u^{,j} = -K_{\mathrm{G}}|\tilde{\nabla}u|^2, \tag{8.19}$$

where K_{G} is the Gaussian curvature, and hence

$$-2gT_{sj}u^{,s}u^{,j} = 2g(K_{\mathrm{G}}|\tilde{\nabla}u|^2 - r^k{}_{,p}u_{,k}u^{,p}), \tag{8.20}$$

where

$$r^k{}_{,p} = \frac{\partial r^k}{\partial x^p} + \Gamma_{pj}{}^k r^j, \qquad \Gamma_{pj}{}^k = \text{Christoffel symbol.}$$

In $N > 2$ dimensions we may use Schwarz's inequality instead of (8.17). This gives

$$\begin{aligned}(P_{,i} - |\tilde{\nabla}u|^2 g'u_{,i} - h'u_{,i})(P^{,i} - |\tilde{\nabla}u|^2 g'u^{,i} - h'u^{,i}) \\ = 4u^{,j}{}_{,i}u^{,i}{}_{,k}u_{,j}u^{,k}g^2 \\ \leq 4u_{,ij}u^{,ij}|\tilde{\nabla}u|^2 g^2.\end{aligned} \tag{8.21}$$

Hence, in $N > 2$ dimensions we have instead of (8.18)

$$\begin{aligned}\tilde{\Delta}P + \frac{L^i P_{,}{}^i}{|\tilde{\nabla}u|^2} \geq & -2g^{-3/2}(g^{-1/2})''|\tilde{\nabla}u|^4 \\ & + \left\{(h' - 2fg)' + \frac{g'}{g}(fg - h')\right\}|\tilde{\nabla}u|^2 \\ & -2gT_{sj}u^{,s}u^{,j} + \frac{h'}{2g}(h' - 2fg),\end{aligned} \tag{8.22}$$

with

$$L^i = \frac{1}{g}(h'u^{,i} - \tfrac{1}{2}P^{,i}) - |\tilde{\nabla}u|^2\left(\frac{g'}{g}u^{,i} - r^i\right).$$

Relations (8.18) and (8.21) are completely analogous to the corresponding ones in the Euclidean case, with the only essential difference that we now have an extra term $-2gT_{sj}u^{,s}u^{,j}$ on the right.

Again, we shall choose g, h such that the corresponding function P satisfies an elliptic inequality. As in Chapter 5, we shall distinguish the two cases in which (a) P assumes its maximum on ∂D, or (b) P assumes its maximum at a critical point of u.

8.2.2 P Attains Its Maximum on ∂D

We consider first a plane domain D, and consequently a two-dimensional Riemannian manifold R^2. First, we prove the analog of Theorem 5.1, which may be stated as

Theorem 8.1 Let u be a sufficiently smooth solution of (8.1.) in a plane domain D, and let $u_m \leq u \leq u_M$. Suppose that the following conditions are satisfied.

(i) The Gaussian curvature K_G associated with the metric tensor $g^{ij} = a^{ij}$ satisfies $K_G \geq K$ in D, and K is finite.

(ii) The following inequality holds for $x \in D$: $r^i{}_{,j}(x)\xi_i\xi^j \le r_0\xi_k\xi^k$, $\boldsymbol{\xi} =$ arbitrary vector.

(iii) For $u_m \le s \le u_M$ we have $g(s) > 0$ and $2(r_0 - K) + f'(s) + 2f(s)(\log g(s))' \le 0$, $(\log g(s))'' \ge 0$

Then, the function

$$P := g(u)a^{ik}u_{,i}u_{,k} + \int_0^u f(y)g(y)\,dy \tag{8.23}$$

assumes its maximum on ∂D.

The proof is just a slight modification of the one for Theorem 5.1. Choosing $h' = fg$ in equation (8.18), we easily check that P satisfies

$$\tilde{\Delta}P - \frac{1}{g}\frac{|\tilde{\nabla}P|^2}{|\tilde{\nabla}u|^2} \ge 0 \qquad \text{in} \quad D$$

if assumptions (i), (ii), (iii) are satisfied, i.e.,

$$\tilde{\Delta}P \ge 0 \qquad \text{in} \quad D.$$

Since $\tilde{\Delta}$ is a uniformly elliptic operator, we may apply the maximum principle, and the statement of Theorem 8.1 follows.

Remarks (a) In the special case when $f(u) \equiv 0$, i.e., u is a solution of

$$Lu = 0 \qquad \text{in} \quad D, \tag{8.24}$$

then condition (iii) reduces to

(iii′) $g(u) > 0$ and $(\log g(u))'' \ge 0$, and P becomes

$$P := g(u)a^{ik}u_{,i}u_{,k}. \tag{8.25}$$

In the special case $a^{ik} = \delta^{ik}$ and $g \equiv 1$ Theorem 8.1 gives the well-known result that the absolute value of the gradient of a harmonic function assumes its maximum on the boundary. On the other hand, Theorem 8.1 also shows that if $(\log g)'' \ge 0$ and

$$\Delta h = 0 \qquad \text{in} \quad D,$$

then the quantity $g(h)|\nabla h|^2$ attains its maximum on ∂D as well.

(b) If u is a solution of

$$\tilde{\Delta}u + f(u) = 0 \qquad \text{in} \quad D, \tag{8.26}$$

where D is a domain on a two-dimensional Riemannian manifold $\mathfrak{M}$, then Theorem 8.1 tells us that

$$P := |\tilde{\nabla}u|^2 g(u) + \int_0^u f(y)g(y)\,dy \tag{8.27}$$

takes its maximum value on ∂D, provided that

(i′) the Gaussian curvature K_G of $\mathfrak{M}$ is nonnegative in D; assumption (iii) holds as in Theorem 8.1 (with $r_0 = 0$).

(c) If u is a solution of

$$\Delta u + \rho(x) f(u) = 0 \qquad \text{in} \quad D \subset E^2, \tag{8.28}$$

with $\rho(x) > 0$ in D, we can interpret u as a solution of (8.26) with

$$\tilde{\Delta} u = \frac{\Delta u}{\rho}, \tag{8.29}$$

and the associated Gaussian curvature becomes

$$K_G = -\frac{\Delta(\log \rho)}{2\rho}. \tag{8.30}$$

(d) If $f(u)$ satisfies

$$(\log f)' \leq 2\beta, \qquad \beta > 0,$$

then assumption (iii) of Theorem 8.1 is satisfied with $g(u) = e^{-\beta u}$.

In $N > 2$ dimensions there is no analog of the identity (8.17). However, the following extension of Lemma 5.4 holds:

Lemma 8.2 For any sufficiently smooth function v the following inequality holds in R^N:

$$N v^{,ik} v_{,ik} \geq (\tilde{\Delta} v)^2. \tag{8.31}$$

Proof Let p^{ij} be a symmetric positive definite tensor, and t_{ik} an arbitrary tensor. Then it is well known that p^{ij} possesses a "square root" Π^{ij}, i.e., $p^{ij} = \Pi^{is}\Pi^{sj}$, and Π^{ij} is symmetric as well. Hence the quadratic form

$$Q := p^{ij} p^{kl} t_{ik} t_{jl}$$

can be written as a sum of squares, namely,

$$Q = \Pi^{is}\Pi^{sj}\Pi^{kr}\Pi^{lr} t_{ik} t_{jl} = q^{sr} q^{sr}$$

with

$$q^{sr} = \Pi^{is}\Pi^{kr} t_{ik}.$$

We now choose

$$p^{ij} = g^{ij}, \qquad t_{ij} = v_{,ij} - \beta \tilde{\Delta} v\, g_{ij}.$$

Then the quadratic form becomes

$$Q = g^{ij} g^{kl} t_{ik} t_{jl} = g^{ij} g^{kl} v_{,ik} v_{,jl} - 2\beta(\tilde{\Delta} v)^2 + N\beta^2(\tilde{\Delta} v)^2,$$

and since $Q \geq 0$ and β is arbitrary, the discriminant condition (or the optimal choice of β) leads directly to the statement of Lemma 8.2.

We are now able to prove

Theorem 8.2 Let u be a sufficiently smooth solution of (8.1) in $D \subset E^N$. Suppose the following conditions are satisfied

(i) the tensor $T_i^j = R_i^j + r^j{}_{,i}$ satisfies for any vector ξ

$$T_i^j \xi^i \xi_j \leq T_0 \xi^k \xi_k \qquad \text{in} \quad D,$$

(ii) the function $g(u)$ satisfies

$$g > 0, \qquad (g^{-1})'' \leq 0$$

and

$$f\left(\frac{2}{N} + 1\right)(\log g)' + 2\left(1 - \frac{1}{N}\right)f' + 2T_0 \leq 0.$$

Then the function

$$P = a^{ik}u_{,i}u_{,k}g(u) + \frac{2}{N}\int_0^u f(y)g(y)\,dy \tag{8.32}$$

assumes its maximum on ∂D.

Proof $h(u)$ has been chosen such that $h' = (2/N)fg$. The combination of Lemma 8.2 and (8.16) leads to

$$\begin{aligned}\tilde{\Delta}P + W^i P_{,i} &\geq -|\tilde{\nabla}u|^4(g^{-1})''g^2 + |\tilde{\nabla}u|^2\left(h'' - fg' - 2f'g - 2\frac{g'h'}{g}\right) \\ &\quad + f\left(h' - \frac{2}{N}fg\right) - 2gT_i^j u_{,j}u^{,i} \\ &\geq -|\tilde{\nabla}u|^4(g^{-1})''g^2 + |\tilde{\nabla}u|^2\left\{\left(\frac{2}{N} - 2\right)f'g - \left(\frac{2}{N} + 1\right)fg'\right. \\ &\quad \left. -2gT_0\right\} \geq 0\end{aligned}$$

because of our assumptions (i) and (ii). The remainder of the proof is as usual.

Remarks (a) Protter and Weinberger (1973) proved a result that is closely related to Theorem 8.2 in the case $f \equiv 0$. They showed, among other things, that for solutions of $Lu = 0$ in D the quantity $|\nabla u|^2$ (ordinary gradient) assumes its maximum on ∂D if the matrix

$$-\frac{\partial b^i}{\partial x^j} - \frac{\partial b^j}{\partial x^i} - \frac{1}{2}(a^{mn})^{-1}\frac{\partial a^{mk}}{\partial x^i}\frac{\partial a^{nk}}{\partial x^j}$$

is positive definite.

(b) An important special case of Theorem 8.2 is

Corollary 8.1 Let u be a solution of (8.1), where $L = \tilde{\Delta}$ is the Laplace–Beltrami operator of a Riemannian space R^N. Suppose that the Ricci tensor of R^N satisfies

$$-R_{ij}\xi^i\xi^j \geq R_0\xi^k\xi_k$$

for some finite R_0, and

$$R_0 \geq \left(1 - \frac{1}{N}\right)f'.$$

Then the function

$$P = |\tilde{\nabla}u|^2 + \frac{2}{N}F(u) \qquad (F' = f)$$

attains its maximum on ∂D.

8.2.3 *P* Assumes Its Maximum at a Critical Point of *u*

The reasoning that may be used here is exactly as in Section 5.4, with some slight and obvious modifications. We shall therefore not repeat all the details of the proofs, but simply point out the necessary changes.

8.2.3.1 Dirichlet Boundary Conditions

Let $\partial/\partial\nu$ denote the intrinsic derivative along a geodesic passing through ∂D in the direction of the exterior normal ν, where ∂D is interpreted as a hypersurface of R^N if $D \subset E^N$. According to Section 8.1, $\partial u/\partial\nu$ corresponds to the conormal derivative $\partial u/\partial N$ with respect to the elliptic operator L (up to a factor that is independent of u). Since u is supposed to vanish on ∂D, we have

$$P = \left(\frac{\partial u}{\partial\nu}\right)^2 g(u) + h(u),$$

so that

$$\frac{\partial P}{\partial\nu} = 2\frac{\partial u}{\partial\nu}\frac{\partial^2 u}{\partial\nu^2}g + \left(\frac{\partial u}{\partial\nu}\right)^3 g'(0) + h'(0)\frac{\partial u}{\partial\nu},$$

where in analogy to the Euclidean case, $\partial^2 u/\partial v^2$ is defined by

$$\frac{\partial^2 u}{\partial v^2} = u_{,ik} v^i v^k,$$

where now the $u_{,ik}$ denote second *covariant* derivatives, and the v^i are contravariant components of the normal v.

By (8.4) and (4.68) we have on ∂D

$$\frac{\partial^2 u}{\partial v^2} + H\frac{\partial u}{\partial v} = -(f(0) + r^i u_{,i}),$$

i.e.,

$$\frac{\partial^2 u}{\partial v^2} = -(f(0) + (Hv^i + r^i)u_{,i}).$$

Here $H = h^\alpha_\alpha$ = trace of the second fundamental tensor of ∂D in the metric $g^{ij} = a^{ij}$. If $h' = 2fg$, we thus find $(\partial u/\partial v = -|\tilde{\nabla} u|)$

$$\frac{\partial P}{\partial v} = \frac{\partial u}{\partial v}(g'(0)|\tilde{\nabla} u|^2 + 2Hg(0)|\tilde{\nabla} u| + 2r^i u_{,i}). \tag{8.33}$$

Let us first consider the case in which D is a plane domain. The result corresponding to Theorem 5.3 is then

Theorem 8.3 Let u be a sufficiently smooth solution of (8.1) in a plane domain D, vanishing on ∂D. Suppose that the following hypotheses are satisfied:

(i) The Gaussian curvature K_G associated with $g^{ij} = a^{ij}$ has the finite lower bound K in D, and for any $x \in D$ the inequality $r^i_{,j}(x)\xi_i\xi^j \le r_0\xi_i\xi^i$ holds.

(ii) $f(u)(\log g(u))' \le 2(K - r_0)$, $\quad g(u)(\log g(u))'' \ge 0$

(iii) With $\tau^2 := \max_{\partial D} a^{ik}u_{,i}u_{,k}$ we have $g'(0)\tau + 2k_0 g(0) - \hat{r}g(0) \ge 0$ where $k_0 = \min_{\partial D} k_g$, k_g = geodesic curvature of ∂D in the metric given by $g_{ik} = (a^{ik})^{-1}$ and $\hat{r} = \max_{\partial D}\sqrt{(a^{ik})^{-1}r^i r^k}$.

Then the function

$$P := a^{ik}u_{,i}u_{,k}g(u) + 2\int_0^u f(y)g(y)\,dy \tag{8.34}$$

assumes its maximum at a point where $\nabla u = 0$.

Proof As before we use the fact that for a two-dimensional manifold we have $R_{is}u^{,i}u^{,s} = -K_G|\tilde{\nabla} u|^2$. Then it is easy to check that under the assumptions (i) and (ii) the right-hand side in (8.18) is nonnegative. On the other hand, if (iii) is satisfied, then by (8.33) we see that

$$\frac{\partial P}{\partial \nu} \leq 0 \qquad \text{on} \quad \partial\Omega.$$

The rest of the proof is analogous to the one for Theorem 5.3.

Remarks (a) Let $\mathfrak{M}$ be a two-dimensional Riemannian manifold, and u the solution of (8.26) with $u = 0$ on ∂D. Then the assumptions of Theorem 8.3. reduce to

(i) $K_G \geq K > -\infty$,
(ii) $f(u)(\log g(u))' \leq 2K$, $g(u)(\log g(u))'' \geq 0$,
(iii) $\tau = \max_{\partial\Omega} |\tilde{\nabla} u|$ and $(g'(0)/(g(0)\tau + 2k_0 \geq 0$.

(b) As in Theorem 5.3, a simple choice of $g(u)$ is $g(u) = e^{-\beta u}$, $\beta > 0$, when assumptions (ii) and (iii) in Theorem 8.3 become

(ii′) $f(u) \geq 2(r_0 - K)/\beta$,
(iii′) $2k_0 - 2\hat{r} \geq \beta\tau$.

In particular, if $r_0 = 0$, $K_G \geq 0$, $k_g \geq k_0 > 0$, and $f(u) \geq 0$, (ii) and (iii′) are satisfied with

$$\beta = 2k_0/\tau.$$

If D is an N-dimensional domain with $N > 2$, the analog of Theorem 8.3 is

Theorem 8.4 Let u be a sufficiently smooth solution of (8.1) in $D \subset E^N$, vanishing on ∂D. Suppose that the following assumptions are satisfied:

(i) The tensor $T_i^j = R_i^j + r_i^j$ satisfies for any $x \in D$ and any vector ξ,

$$T_i^j(x)\xi_j\xi^i \leq T_0\xi_k\xi^k, \qquad T_0 \text{ finite;}$$

(ii) $g(u) > 0$, $(g^{-1/2}(u))'' \leq 0$ and

$$g'(u)f(u) + 2g(u)T_0 \leq 0.$$

(iii) The average curvature H of ∂D (in the metric with $g^{ij} = a^{ij}$) is bounded below by H_0 and

$$g'(0)\tau + 2H_0 g(0) - 2\hat{r}g(0) > 0,$$

(τ and $\hat{r}$ as in Theorem 8.3).

Then the function

$$P := a^{ik}u_{,i}u_{,k}g(u) + 2\int_0^u f(y)g(y)\,dy \tag{8.35}$$

assumes its maximum where $\nabla u = 0$.

Proof Note that $T_{ik}u^{,i}u^{,k} = T_i^j u^{,i}u_{,j} \leq T_0|\tilde{\nabla} u|^2$ by assumption (i). Together with our hypothesis (ii) it follows that the quadratic form in $|\tilde{\nabla} u|^2$ in inequality (8.22) is positive definite. Again, $\partial P/\partial \nu \leq 0$ on ∂D if (iii) holds. The rest of the proof consists of the usual arguments.

Remarks (a) It follows from Theorem 8.4 that the function

$$P := a^{ik}u_{,i}u_{,k} + 2\int_0^u f(y)\,dy$$

assumes its maximum at a critical point of u, provided that

(i′) $T^i_j(x)$ is negative semidefinite for all $x \in D$,
(ii′) $r^i \nu_i + H \geq 0$ on ∂D,

where $\nu_i = a_{is}n^s$, $a_{is} = (a^{is})^{-1}$, and $\mathbf{n}$ is the outward normal on ∂D. This special version of Theorem 8.4. was given in Sperb (1979a).

(b) If u is a solution of (8.28) in $D \subset E^N$, one finds after some calculations that conditions (i′) and (ii′) are satisfied provided the "inhomogeneity" $\rho(x)$ satisfies

(i″) $\Delta(\rho^{-1/2}(x)) \leq 0$, $\rho(x) > 0$, $x \in D$,
(ii′) $H = (N-1)K_a + \frac{1}{2}(\partial/\partial n)(\log \rho) > 0$ on ∂D and K_a is the average curvature of ∂D in the Euclidean sense.

This special case is treated in the paper of Schaefer and Sperb (1977). Note that P defined by (8.35) then takes the form

$$P = \frac{|\nabla u|^2}{\rho} + 2\int_0^u f(y)\,dy.$$

8.2.3.2 Neumann Boundary Conditions

In the case for which the solution of (8.1) satisfies the boundary condition

$$\frac{\partial u}{\partial N} = 0 \qquad \text{(conormal derivative)} \qquad \text{on} \quad \partial D \subset E^N, \tag{8.36}$$

which is equivalent to

$$\frac{\partial u}{\partial \nu} = 0 \qquad \text{on} \quad \partial D \subset R^N, \tag{8.37}$$

there is no need for a new analysis now. We can follow the reasoning of Section 5.4.2, the only difference being that instead of considering ∂D as a hypersurface in E^N, we have to consider ∂D as a hypersurface in R^N, with the metric in R^N given by $g^{ij} = a^{ij}$, so that the metric tensor $g_{\alpha\beta}$ of $\partial D \subset R^N$ is given by

$$g_{\alpha\beta} = g_{ij}\frac{\partial x^i}{\partial y^\alpha}\frac{\partial x^j}{\partial y^\beta}, \tag{8.38}$$

where ∂D is represented by $x^i = x^i(y^\alpha)$. Instead of the Euclidean normal n to ∂D we have to use $v^i = a^{ik}n_k$ again and use generalized covariant derivatives in the place of covariant derivatives. Exactly the same calculations as the ones leading to (5.44) now show that if u satisfies Neumann boundary conditions, then

$$\frac{\partial P}{\partial v} = -2g(u) \sum_{\alpha=1}^{N-1} \kappa_\alpha u_{,\alpha} u^{,\alpha}, \tag{8.39}$$

where the κ_α are the eigenvalues of h_β^γ, i.e., the mixed second fundamental form of $\partial D \subset R^N$. In view of (8.39) one is led to the result

Theorem 8.5 Let u be a sufficiently smooth solution of (8.1) that satisfies Neumann boundary conditions (8.36). Suppose that the conditions (i), (ii) of Theorem 8.3 for $N = 2$ or (i), (ii) of Theorem 8.4, for $N > 2$, are satisfied and that

(iii) the second fundamental tensor h_α^β of $\partial D \subset R^N$ is positive semidefinite (i.e., all $\kappa_\alpha \geq 0$).

Then the function

$$P := a^{ik} u_{,i} u_{,k} g(u) + 2 \int_0^u f(y) g(y)\, dy$$

assumes its maximum where $\nabla u = 0$.

Remarks (a) If u is a solution of (8.28) that satisfies Neumann boundary conditions, then assumption (iii) of Theorem 8.5 is satisfied, provided that

$$k_{\min} + \frac{1}{2} \frac{\partial}{\partial n} (\log \rho) \geq 0 \qquad \text{on} \quad \partial D, \tag{8.40}$$

where $k_{\min}$ denotes the minimal normal curvature of $\partial D \subset E^N$.

(b) Of course, Corollary 5.3 carries over to the "Riemannian case" with the obvious modifications.

8.2.3.3 Nonlinear Boundary Conditions in the Plane

As in Chapter 5 we have to restrict our attention to plane domains D if we consider boundary conditions of the form

$$\gamma \frac{\partial u}{\partial N} + \sigma(u) = \frac{\partial u}{\partial v} + \sigma(u) = 0 \qquad \text{on} \quad \partial D. \tag{8.41}$$

All calculations of Section 5.4.3 can be repeated now with slight and obvious modifications.

(i) $\partial/\partial n$ is to be replaced by $\partial/\partial v$,
(ii) $\partial/\partial s$ denotes an intrinsic derivative in ∂D,

(iii) the curvature k has to be replaced by the geodesic curvature k_g (in the induced metric in ∂D!).

Again, we need only give sufficient conditions ensuring that

(a) $P = |\nabla u|^2 g(u) + h(u)$ satisfies an elliptic differential inequality, and
(b) $\partial P/\partial \nu \leq 0$ holds on ∂D.

The next result is then the "elliptic analog" of Theorem 5.7.

Theorem 8.6 Let u be a sufficiently smooth solution of (8.1) satisfying the boundary condition (8.41) on ∂D, where D is a plane domain. Suppose that in addition to hypotheses (i) and (ii) of Theorem 8.3, the following assumptions hold:

(iii) The geodesic curvature k_g of ∂D in the metric induced by $g_{ij} = (a^{ij})^{-1}$ is bounded below by k_0, and we have

$$2k_0 + \sigma(\log g)' \geq 0,$$
$$2(\sigma\sigma' + f) + \sigma^2(\log g)' \geq 0,$$
$$(\log g)' + \frac{k_0}{\sigma} + (\log \sigma)' \geq 0.$$

Then, the function

$$P = a^{ik}u_{,i}u_{,k}g(u) + 2\int_0^u f(y)g(y)\,dy$$

assumes its maximum at a point where $\nabla u = 0$.

Remarks (a) The analog of Corollary 5.4 of Theorem 5.7 holds again here with the obvious changes (i.e., "geodesically convex" instead of "convex").

(b) For a general elliptic operator, it is rather difficult to verify whether or not all assumptions, say of Theorem 8.6, are satisfied. In many important cases, however, the elliptic operator L under consideration *is* the Laplace–Beltrami operator of a surface.

8.3 APPLICATIONS

8.3.1 Isoperimetric Inequalities in the Torsion Problem on a Manifold

Let $\mathfrak{M}$ be a two-dimensional Riemannian manifold and D a finite, bounded domain. We then consider the problem (Sperb, (1980b))

$$\tilde{\Delta} u + 1 = 0 \qquad \text{in} \quad D \tag{8.42}$$

with Dirichlet data on ∂D. The next result is an immediate consequence of Theorem 8.1.

Corollary 8.2 Let u be a $C^3(D) \cap C^2(\bar{D})$ solution of (8.42). Suppose the Gaussian curvature K_G of $\mathfrak{M}$ satisfies $K_G \geq K > -\infty$ in D. Then, the function

$$P = \left(|\tilde{\nabla} u|^2 + \frac{1}{K}\right) e^{Ku} \tag{8.43}$$

attains its maximum on ∂D.

Proof Since $r_0 = 0$, $g = e^{Ku}$, $h = (1/K)e^{Ku}$, and $f \equiv 1$, the function P given by (8.43) obviously satisfies the hypotheses of Theorem 8.1, which implies Corollary 8.2.

Let τ now be defined as

$$\tau = \max_{\partial D} |\tilde{\nabla} u|, \tag{8.44}$$

u being a solution of (8.42). The following result is a consequence of Corollary 8.2.

Theorem 8.7 Let u be a solution of (8.42) vanishing on ∂D, and suppose that $k_g \geq k$ and $K_G \geq K$. Then,

$$\tau \leq \sqrt{(k/K)^2 + (1/K)} - k/K \qquad \text{if} \quad K > 0, \tag{8.45}$$

$$\tau \geq \sqrt{(k/K)^2 + (1/K)} - k/K \qquad \text{if} \quad K < 0. \tag{8.46}$$

Proof Since $u = 0$ on ∂D, we have $\partial u/\partial v = -|\tilde{\nabla} u|$. P assumes its maximum on ∂D, say at x, and therefore we have at x

$$\frac{\partial P}{\partial v} = K \frac{\partial u}{\partial v}\left(|\tilde{\nabla} u|^2 + \frac{1}{K}\right) + 2\frac{\partial u}{\partial v}\left(-1 - k_g \frac{\partial u}{\partial v}\right) > 0$$

for any nonconstant P, which implies that at x

$$2k_g\tau + K\tau^2 < 1, \tag{8.47}$$

which in turn leads to (8.45) or (8.46) respectively. Note that inequality (8.47) is nontrivial only if k and K are not both negative.

Remarks (a) It seems worthwhile to mention the following special case. Let D be the interior of a geodesic circle C_g on a sphere, i.e.,

$$C_g\colon \quad \{r = R, \quad 0 \leq \phi \leq 2\pi, \quad 0 \leq \theta \leq \theta_0\},$$

where r, ϕ, θ are spherical coordinates. Then it is not hard to see that

$$u = u(\theta) = R^2 \log\left(\frac{1 + \cos\theta}{1 + \cos\theta_0}\right)$$

is the solution of (8.42), vanishing for $\theta = \theta_0$. Furthermore,

$$P = R^2/(1 + \cos\theta_0)^2 = \text{const}$$

in this case, and (8.47) and (8.45) become equalities.

(b) Since

$$A = \oint_{\partial D} \left(-\frac{\partial u}{\partial \nu}\right) ds \le \tau L,$$

where A is the area of D and L the length of ∂D, it follows also from (8.47) that $|\tilde{\nabla} u|$ cannot assume its maximum τ on a part of ∂D where

$$k_g \ge \frac{L}{2A} - K\frac{A}{L}. \tag{8.48}$$

(c) Note that no assumption on k_g and K_G (except for finiteness) had to be made. If we then use the inequality $\tau \ge A/L$ in (8.47), we get the purely geometric inequality

$$2k_g(x)\frac{A}{L} + K\frac{A^2}{L^2} < 1,$$

i.e.,

$$2k\frac{A}{L} + K\frac{A^2}{L^2} \le 1, \tag{8.49}$$

with equality if D is a geodesic disk. Note that the Gauss–Bonnet theorem gives

$$2\pi + \int_D K_G = \oint_{\partial D} k_g\, ds \ge kL,$$

so that

$$2\pi + K_M A \ge kL, \qquad K_M = \max_D K_G,$$

whereas we get

$$2kL + KA \le \frac{L^2}{A}.$$

Corollary 8.2, stated as an inequality, reads

$$\left(|\tilde{\nabla} u|^2 + \frac{1}{K}\right) e^{Ku} \le \tau^2 + \frac{1}{K},$$

which becomes, at the point in D where $u = u_M := \max_D u$,

$$e^{Ku_M} \le K\tau^2 + 1, \tag{8.50}$$

Combined with (8.45), it gives

$$u_M \leq \frac{1}{K} \log\left(\frac{2}{K}(\alpha - k\sqrt{\alpha})\right), \qquad \alpha := K + k^2 \tag{8.51}$$

if $K > 0$. For $K < 0$ the corresponding inequalities cannot be combined so that no counterpart of (8.51) exists in that case.

Up to this point we have only used a function P, taking its maximum on ∂D. We shall now see how a minor change in P will alter its character, as the next result illustrates.

Corollary 8.3 Let u be a smooth solution of (8.42), vanishing on ∂D. Suppose that $k_g \geq 0$ and $K_G \geq K > 0$ in D. Then the function

$$P := \left(|\tilde{\nabla} u|^2 + \frac{1}{K}\right) e^{2Ku} \tag{8.52}$$

assumes its maximum where $\nabla u = 0$.

Proof We apply Theorem 8.3 noting that $r_0 = 0$ and that the inequalities in assumption (ii) there become equalities with our present choice of $g(u)$, $h(u)$. Clearly, hypothesis (iii) is satisfied as well, and thus Theorem 8.3 implies Corollary 8.3.

Remarks Actually, hypothesis (iii) in Theorem 8.3 would be satisfied if

$$2K\tau + 2k \geq 0 \qquad \text{on} \quad \partial D, \tag{8.53}$$

where $K_G \geq K$, $k_g \geq k$, but no assumption on the signs of k and K are made. For instance, if $K > 0$, (8.53) would be satisfied provided that

$$2K\frac{A}{L} + 2k > 0,$$

so that k_g is allowed to become negative.

Before studying some applications of Corollary 8.3, we note that the optimal domain for Corollary 8.3 now is a "geodesic strip" S_g given by

$$S_g = \{r = R, 0 \leq \phi \leq 2\pi, \qquad \tfrac{1}{2}\pi - \alpha \leq \theta \leq \tfrac{1}{2}\pi + \alpha, 0 \leq \alpha < \tfrac{1}{2}\pi\},$$

if $D = S_g$, the solution of (8.42), is

$$u = R^2 \log\left(\frac{\sin\theta}{\cos\alpha}\right),$$

while P, given by (8.52), becomes

$$P = \frac{R^2}{\cos^2\alpha} = \text{const.}$$

Hence all inequalities deduced from Corollary 8.3 will be equalities if $D = S_g$. On ∂D Corollary 8.3 implies

$$\left(|\tilde{\nabla} u|^2 + \frac{1}{K}\right) \le \frac{1}{K} e^{2Ku_M},$$

and hence

$$\tau^2 \le \frac{1}{K}(e^{2Ku_M} - 1). \tag{8.54}$$

We can continue (8.54) on the left since $\tau \ge A/L$, to get

$$u_M \ge \frac{1}{2K} \log\left(\frac{KA^2}{L^2} + 1\right). \tag{8.55}$$

We now derive an upper bound for u_M in the usual way. Let Q be a point in D where $u = u_M$ and R a point on ∂D nearest to Q in geodesic distance. Let s measure the distance from Q along the geodesic joining Q and R. Clearly, we have

$$-\frac{du}{ds} \le |\tilde{\nabla} u|,$$

and hence by Corollary 8.3

$$\int_0^{u_M} (e^{2K(u_M - u)} - 1)^{-1/2}\, du \le d\sqrt{K}, \tag{8.56}$$

where d is the geodesic distance of Q and R. Inequality (8.56) then leads to

$$u_M \le \frac{1}{K} \log \frac{1}{\cos(d\sqrt{K})}, \tag{8.57}$$

with equality again if $D = S_g$. Of course, we can take for d the radius of the largest geodesic circle inscribed in D. The combination of (8.54) and (8.57) finally gives

$$\tau^2 \le \frac{1}{K} \frac{1}{\cos^2(d\sqrt{K})} - 1 = \frac{1}{K} \tan^2(d\sqrt{K}). \tag{8.58}$$

Remarks (a) The inequalities (8.55) and (8.57) again imply a purely geometric inequality, namely

$$\tan^2(d\sqrt{K}) \ge \frac{KA^2}{L^2}, \tag{8.59}$$

which is valid under the assumptions of Corollary 8.3 and again with S_g as the optimal domain.

(b) Of course, one could also combine the inequalities obtained from Corollary 8.2 with those derived from Corollary 8.3. However, since the optimal domains are different in the two cases, the resulting inequalities will not be isoperimetric.

(c) It seems worthwhile to mention that the results given in this section may be applied to the solution of the Poisson problem, i.e., the solution of

$$\Delta u = -\rho(x) \quad \text{in} \quad D \subset E^2, \qquad \rho > 0, \tag{8.60}$$

vanishing on ∂D. All inequalities in this section may be applied when using the following "translation formulas":

$$\tau := \max_{\partial D} \frac{|\nabla u|}{\rho},$$

$$K_G = -\frac{\Delta(\log \rho)}{2\rho}, \qquad \text{i.e.,} \quad K := -\max_{D} \frac{\Delta(\log \rho)}{2\rho},$$

$$k_g = \kappa + \frac{1}{2}\frac{\partial}{\partial n}(\log \rho), \qquad \kappa \text{ is the ordinary curvature,}$$

$$k = \min_{\partial D} k_g,$$

$$A = \int_D \rho \, dx, \qquad L = \oint_{\partial D} \sqrt{\rho} \, ds.$$

8.3.2 Eigenvalue Problems on a Manifold

In various contexts one is interested in the eigenvalue problem

$$\tilde{\Delta} u + \lambda u = 0 \quad \text{in} \quad D, \tag{8.61}$$

complemented either with some boundary conditions (mostly Dirichlet or Neumann) if D is a subdomain of an N-dimensional Riemannian manifold, or periodicity conditions if $D = \mathfrak{M}$, where $\mathfrak{M}$ is a finite manifold without boundary (e.g., the surface of a sphere). In the latter case the first nonzero eigenvalue λ_1 is also called the "first eigenvalue of $\mathfrak{M}$."

In recent years there has been an increasing interest in obtaining a priori bounds for the first eigenvalue λ_1 of a manifold, mainly from a purely geometric point of view. But there are also interesting aspects in other fields, as for example, illustrated in the paper of Murray and Sperb (1981a) where a priori bounds for λ are of importance in the context of a biological model.

We shall first study applications of maximum principles similar to those derived in the beginning of this chapter. Of main interest is the eigenvalue problem (8.61), but it is evident that other applications to problems on a manifold are possible as well.

8.3.2.1 Bounds for the First Eigenvalue of a Manifold

It is convenient to state a simple auxiliary result first.

Lemma 8.3 Let v be any C^2 function satisfying

$$\tilde{\Delta} v + b^i v_{,i} \geq 0$$

on a compact manifold $\mathfrak{M}$ without boundary, where the b^i are bounded functions on $\mathfrak{M}$. Then $v \equiv \text{const}$.

Proof Suppose $v \not\equiv \text{const}$. Then v must assume its maximum at some point $x \in \mathfrak{M}$. Let Ω be an arbitrary subdomain of $\mathfrak{M}$ containing x in its interior. However, by the maximum principle $x \in \partial\Omega$ must hold unless $v \equiv \text{const}$. Thus the assumption on $v \not\equiv \text{const}$ leads to a contradiction.

An immediate consequence of Lemma 8.3 and a calculation similar to the one leading to Theorem 8.2 show that the following result holds.

Theorem 8.8 Suppose that $\mathfrak{M}$ is a compact, N-dimensional C^2-manifold without boundary, whose Ricci curvature is bounded below by $(N-1)K$, $k > 0$. Then

$$\lambda_1 \geq NK. \tag{8.62}$$

Proof The Ricci curvature is defined here as the lowest eigenvalue of $-R_{ij}$, R_{ij} being the Ricci tensor of $\mathfrak{M}$. We then choose

$$P = |\tilde{\nabla} u|^2 + \frac{\lambda_1}{N} u^2,$$

u being the first nonconstant eigenfunction of $\mathfrak{M}$. The calculations that proved Theorem 8.2 then show that P satisfies

$$\tilde{\Delta} P \geq |\tilde{\nabla} u|^2 \lambda_1 \left(\frac{2}{N} - 2\right) - R^s_j u_{,s} u^{,j} \geq 2(N-1)|\tilde{\nabla} u|^2 \left(-\frac{\lambda_1}{N} + K\right).$$

According to Lemma 8.3, the last expression in brackets on the right cannot be positive throughout $\mathfrak{M}$ unless $P \equiv \text{const}$, i.e., when $\mathfrak{M}$ is an N-sphere. This proves Theorem 8.8.

Remark A different proof of Theorem 8.8, based on a result of Lichnerowicz, is given in the volume by Berger *et al.* (1971).

An essential feature of Lemma 8.3 is the fact that the coefficients b^i are bounded in $\mathfrak{M}$. Another simple auxiliary result treats the case in which at least one of the b^i becomes unbounded, say at the point x_0.

Lemma 8.4 Suppose the C^2 function v satisfies

$$\tilde{\Delta} v + b^i v_{,i} \geq 0 \quad (\leq 0) \qquad \text{in} \quad \mathfrak{M},$$

where $\mathfrak{M}$ is a compact manifold without boundary, and the b^i are bounded except at the point x_0. Then v must assume its maximum (minimum) in $\mathfrak{M}$ at x_0.

Proof Set $M_0 = \mathfrak{M} - N(x_0)$, where $N(x_0)$ is some neighborhood of x_0. By the maximum principle v must attain its maximum (minimum) on $\partial M_0 = \partial N_0$. Letting $N(x_0)$ shrink to $\{x_0\}$, we are led to the statement of Lemma 8.4.

We shall say that an eigenfunction u is symmetric if $u_M := \max_{\mathfrak{M}} u = -\min_{\mathfrak{M}} u = -u_m$. For example, if $\mathfrak{M}$ is realized as a surface Σ in space, u will be symmetric if Σ has two perpendicular planes of symmetry.

As a consequence of inequality (8.22) and Lemma 8.4. we have

Theorem 8.9 Suppose $\mathfrak{M}$ is a compact manifold without boundary, whose Ricci curvature is nonnegative. Let d be the diameter of $\mathfrak{M}$. Then

$$\lambda_1 \geq \frac{\pi^2}{4d^2}, \tag{8.63}$$

and if u is symmetric, we even have

$$\lambda_1 \geq \frac{\pi^2}{d^2}, \tag{8.64}$$

Proof According to inequality (8.22), the function

$$P = |\tilde{\nabla} u|^2 + \lambda_1 u^2$$

satisfies

$$\tilde{\Delta} P + \frac{L^i P_{,i}}{|\tilde{\nabla} u|^2} \geq -2R^s_j u_{,s} u^{,j} \geq 0 \qquad \text{in} \quad \mathfrak{M}.$$

By Lemma 8.4 P must assume its maximum at a point where $|\tilde{\nabla} u| = 0$. We can normalize u such that $u_M = 1$ and $u_m \geq -1$. Lemma 8.4 thus tells us that

$$|\tilde{\nabla} u|^2 + \lambda_1 u^2 \leq \lambda_1 \qquad \text{in} \quad \mathfrak{M}.$$

Let γ be a geodesic joining a point x_M where $u = u_M = 1$ to a point x_m where $u = u_m \geq -1$. Denote by s the arc length along γ. Since

$$\left|\frac{du}{ds}\right| \leq |\nabla u| \leq \sqrt{\lambda_1}\sqrt{1 - u^2},$$

we have

$$\int_{u_m}^{1} \frac{du}{\sqrt{1 - u^2}} \leq \sqrt{\lambda_1}\,\mathrm{dist}(x_m, x_M) \leq \sqrt{\lambda_1}\, d.$$

If u is a symmetric, we have $u_m = -1$. If u is not symmetric we may replace u_m by 0. This leads to the two bounds given in Theorem 8.9.

Remarks (a) It was shown by Cheng (1975) that if the Gaussian curvature K_G of $\mathfrak{M}$ is nonnegative (in the two- dimensional case), one has

$$\lambda_1 \geq \frac{1}{4d^2}.$$

Note that (8.63) is an improvement of this latter inequality by a factor 10 roughly.

(b) If the Ricci curvature of $\mathfrak{M}$ has the negative lower bound $-(N-1)K$, $K > 0$, then one easily checks that the function

$$P = |\tilde{\nabla} u|^2 + (\lambda_1 + (N-1)K)u^2$$

also assumes its maximum where $|\tilde{\nabla} u| = 0$. Reasoning similar to that given above then shows that

$$\lambda_1 \geq \frac{\pi^2}{4d^2} - (N-1)K$$

or

$$\lambda_1 \geq \frac{\pi^2}{d^2} - (N-1)K \qquad (u \text{ symmetric})$$

holds. Both bounds are of course only interesting if the quantities on the right are positive.

(c) Suppose that the two-dimensional manifold $\mathfrak{M}$ has a strictly positive Gaussian curvature, i.e.,

$$K_G \geq K > 0.$$

Suppose furthermore that the first nonconstant eigenfunction is symmetric. We then use Eq. (8.18) and choose $g(u) = e^{-\beta u}$ and $h(u) = 2\lambda_1 \int_0^u ve^{-\beta v}\,dv$. The quadratic form in $|\tilde{\nabla} u|^2$ in (8.18) will then be positive definite, provided that

$$\beta\lambda_1 u + 2K \geq 0.$$

We normalize u such that $u_m = -1$, $u_M = 1$. Then our last inequality is satisfied if we choose

$$\beta = \frac{2k}{\lambda_1}.$$

Integration analogous to that leading to (8.64) now yields

$$\int_{-1}^{1} \left(\int_u^1 e^{-\beta(y-u)} y\,dy \right)^{-1/2} du \leq \sqrt{2\lambda_1}\,d, \qquad \beta = \frac{2K}{\lambda_1}. \tag{8.65}$$

Since (8.65) gives a rather involved inequality for λ_1, one may approximate the left-hand side in (8.65) by a series expansion. This leads to a somewhat simpler inequality. For $x = 1/\sqrt{\lambda_1}$ one has

$$\pi x + \tfrac{2}{3}Kx^3 \leq d. \tag{8.66}$$

Note that when $\mathfrak{M}$ is a two-sphere of radius 1, one has $\lambda_1 = 2$. Inequality (8.64) yields $\lambda_1 \geq 1$, whereas (8.66) gives approximately $\lambda_1 \geq 1.34$.

8.3.2.2 Eigenvalues of a Subdomain $D \subset \mathfrak{M}$

Let D be a finite subdomain in a manifold $\mathfrak{M}$. First, we consider the problem of a "vibrating membrane on a manifold," i.e., the problem

$$\tilde{\Delta} u + \nu u = 0 \quad \text{in} \quad D, \qquad u = 0 \quad \text{on} \quad \partial D. \tag{8.67}$$

Our next result is the analog of Theorem 8.8 in the case of problem (8.67).

Theorem 8.10 Assume that the Ricci curvature of $\mathfrak{M}$ has the positive lower bound $(N-1)K$ in D and that the average curvature of ∂D (i.e., trace of the second fundamental form of ∂D) is nonnegative. Then the following inequality holds:

$$\nu_1 \geq NK. \tag{8.68}$$

Proof As in the proof of Theorem 8.8, we consider the function

$$P = |\tilde{\nabla} u|^2 + \frac{\nu_1}{N} u^2,$$

u being the first eigenfunction of (8.67). We can now apply Theorem 8.2, noting that $T_i^j = R_i^j$, $T_0 = -(N-1)K$, and $g \equiv 1$ in our case. The hypotheses of Theorem 8.2 are satisfied if

$$\left(1 - \frac{1}{N}\right)\nu_1 - (N-1)K \leq 0,$$

i.e., if

$$\nu_1 < NK$$

holds. Let us suppose that this last inequality holds. Theorem 8.2 then tells us that P must assume its maximum on ∂D. At a point of ∂D where P assumes its maximum we must have $\partial P/\partial \boldsymbol{\nu} > 0$ ($\boldsymbol{\nu}$ is the normal to ∂D here). But on ∂D, we have

$$\begin{aligned} \frac{\partial P}{\partial \upsilon} &= \frac{\partial}{\partial \upsilon}\left(\frac{\partial u}{\partial \nu}\right)^2 + \frac{2}{N}\nu_1 u \frac{\partial u}{\partial \nu} \\ &= 2\frac{\partial^2 u}{\partial \nu^2}\frac{\partial u}{\partial \nu} = -2(N-1)H\left(\frac{\partial u}{\partial \nu}\right)^2 \leq 0 \end{aligned}$$

since we have assumed that the average curvature H of ∂D is nonnegative. Hence $v < NK$ is impossible, and therefore (8.68) must hold.

Remarks (a) The equality sign holds in (8.68) if $\mathfrak{M}$ is a sphere and ∂D a great circle.

(b) It was shown by McKean (1970), that if the Gaussian curvature is bounded above by a negative quantity, say $-\alpha^2$, then one has $v_1 \geq \frac{1}{4}\alpha^2$, and a similar inequality holds in N dimensions.

There is not much that needs to be altered in the proof of Theorem 8.10 if, instead of problem (8.67), we consider the "free membrane problem on $\mathfrak{M}$," namely,

$$\tilde{\Delta} u + \mu u = 0 \quad \text{in} \quad D, \qquad \frac{\partial u}{\partial v} = 0 \quad \text{on} \quad \partial D. \tag{8.69}$$

We now have

Theorem 8.11 Assume that the Ricci curvature of $\mathfrak{M}$ has the positive lower bound $(N-1)K$ in D, and ∂D is convex, i.e., the second fundamental form of ∂D is positive semidefinite. Then the first positive eigenvalue μ_1 of (8.69) satisfies

$$\mu_1 \geq NK. \tag{8.70}$$

Proof We proceed exactly as in the proof of Theorem 8.10. It remains to check that $\partial P/\partial v \leq 0$ would hold on ∂D. But, by equation (8.39) we get

$$\frac{\partial P}{\partial v} = -2 \sum_{\alpha=1}^{N-1} \kappa_\alpha u_{,\alpha} u^{,\alpha} \leq 0$$

since by assumption all eigenvalues κ_α of the second fundamental form are nonnegative. This proves (8.70).

Of course, there are also analogs of Theorem 8.9 in the two eigenvalue problems (8.67) and (8.69). For the fixed membrane problem we have

Theorem 8.12 Suppose the Ricci curvature of $\mathfrak{M}$ is nonnegative in D and the average curvature of ∂D is nonnegative as well. Then

$$v_1 \geq \frac{\pi^2}{4\rho^2}, \tag{8.71}$$

where ρ is the radius of the largest geodesic ball contained in D.

Proof The function $P = |\tilde{\nabla} u|^2 + v_1 u^2$ assumes its maximum at a critical point of u. This follows as a special case of Theorem 8.4. Let x_M be a point in D where u assumes its maximum, and let x_0 be a point on ∂D nearest to

x_M in geodesic distance. Then the usual integration gives (choose $\max_D u = 1$)

$$\frac{\pi}{2} = \int_0^1 \frac{du}{\sqrt{1-u^2}} \le \sqrt{v_1}\,\mathrm{dist}(x_M, x_0) \le \sqrt{v_1}\rho,$$

which is equivalent to inequality (8.71).

Remarks (a) For a two-dimensional manifold it was shown (see Osserman (1978)) that if

$$\int_D K^+\,dA < 2\pi, \qquad K^+ = \max\{K_G, 0\},$$

then one has

$$v_1 \ge \frac{1}{4\rho^2}.$$

(b) If the Ricci curvature is not nonnegative everywhere, but still has a finite, negative lower bound $-(N-1)K$, $K > 0$, then one can choose $P = |\tilde{\nabla} u|^2 + (\lambda_1 + (N-1)K)u^2$. The same reasoning as above then shows that

$$v_1 \ge \frac{\pi^2}{4\rho^2} - (N-1)K$$

must hold if ∂D has nonnegative average curvature.

(c) In the case in which D is a plane domain (and $\tilde{\Delta} = \Delta$), the inequality (8.71) was derived by Hersch (1960), who employed different methods.

Finally, there is a similar lower bound as (8.71) for the eigenvalue μ_1. We state the result as

Theorem 8.13 Suppose the Ricci curvature of $\mathfrak{M}$ is nonnegative in D and the second fundamental form of ∂D is positive semidefinite (for $N = 2$, ∂D geodesically convex). Then

$$\mu_1 \ge \frac{\pi^2}{4d^2}, \tag{8.72}$$

and if the first eigenfunction is symmetric one even has

$$\mu_1 \ge \frac{\pi^2}{d^2}, \tag{8.73}$$

where d is the diameter of D.

Proof Under our assumptions, the function $P = |\tilde{\nabla} u|^2 + \mu_1 u^2$ attains its maximum where $|\tilde{\nabla} u| = 0$. This follows directly from Theorem 8.5. Inequalities (8.72) and (8.73) then follow with the same arguments as those in the proof of Theorem 8.9.

Remarks (a) For a two-dimensional manifold inequality (8.73) was proven with entirely different methods by Chavel and Feldman (1977) under the same assumptions as in Theorem 8.13, except for the symmetry assumption, which they did not need. In view of their result, it seems probable that (8.73) holds without the assumption that the eigenfunction be symmetric. On the other hand, if D is symmetric, e.g., a domain on a surface in space possessing two perpendicular planes of symmetry, one can replace the diameter d by a better constant, similar to that in inequality (6.45).

(b) One could also derive an extension of inequality (6.29) to the case of the problem

$$\Delta u + \lambda(\alpha)u = 0 \quad \text{in} \quad D, \qquad \frac{\partial u}{\partial \nu} + \alpha u = 0 \quad \text{on} \quad \partial D, \tag{8.74}$$

using arguments similar to those used in the Euclidean case. Since the reasoning is somewhat involved, it is left to any interested reader.

(c) Of course, numerous other bounds in all eigenvalue problems on a manifold could be derived by various integrations of the inequalities resulting from maximum principles. For example, under the hypotheses of Theorem 8.12, one has

$$|\tilde{\nabla} u|^2 + \nu_1 u^2 \leq \nu_1 u_M^2 \quad \text{in} \quad D.$$

An integration over D and application of the generalized Green's identity therefore yields

$$\int_D u^2\, dx \leq \frac{A}{2}\, u_M^2,$$

where $u_M = \max_D u$ and A is the "N-area" of D.

Chapter 9

MAXIMUM PRINCIPLES ASSOCIATED WITH SOLUTIONS OF $u_{,t} = \Delta u + f(u)$

Numerous important models in chemistry and biology (see Chapter 1) lead to a nonlinear reaction-diffusion equation of the form

$$u_{,t} = \delta\,\Delta u + f(u) \qquad \text{in} \quad \Omega \times (0, T), \tag{9.1}$$

where δ is the (positive) diffusion coefficient and Ω a domain in E^N. Often the variables t and x are chosen such that δ becomes a dimensionless constant, which then will be chosen to be 1. Sometimes, however, it is convenient to introduce a parameter λ and write instead of (9.1)

$$u_{,t} = \Delta u + \lambda f(u) \qquad (\text{in} \quad \Omega \times (0, T). \tag{9.2}$$

Most of the results of this chapter could easily be generalized to the case in which instead of the cross product $\Omega \times (0, T)$ one has a general domain $D \subset \mathbb{R}^N \times \mathbb{R}^+$, where the Laplacian is replaced by a uniformly elliptic operator. But in view of the applications, we shall restrict ourselves to the simpler case in which $D = \Omega \times (0, T)$, $0 < T < \infty$. For $\delta = 0$ (9.1) becomes the ordinary differential equation

$$\dot{z} = f(z) \tag{9.3}$$

for the function $z(t)$, which can be integrated immediately in the form

$$\int_{z(0)}^{z(t)} \frac{dy}{f(y)} = t. \tag{9.4}$$

On the other hand, (9.1) and (9.2) also have the time-independent solutions $u(x)$ of

$$\Delta u + \lambda f(u) = 0 \quad \text{in} \quad \Omega \tag{9.5}$$

as possible steady-states. The solutions of (9.1) and (9.2) will have many features in common with both the solutions of (9.3) and (9.5). Therefore we shall associate different types of maximum principles with the solutions of (9.1) and (9.2). Some of them will display features similar to the solutions of (9.3), and some will be closer to solutions of (9.5).

9.1 MAXIMUM PRINCIPLES FOR FUNCTIONALS OF THE FORM $v = g(u)u_{,t} + h(u)$

The "Ansatz" chosen in the title of this section is related to the function P used in the previous chapters. We shall see that it is possible to choose g and h such that v satisfies a parabolic maximum principle. From this a number of useful applications will follow

Let u now be a solution of the initial boundary-value problem

$$\begin{aligned} u_{,t} &= \Delta u + f(u) && \text{in} \quad \Omega \times (0, T), \\ u &= 0 && \text{on} \quad \partial\Omega \times (0, T), \\ u(x, 0) &= u_0(x). \end{aligned} \tag{9.6}$$

Our first result is then (see Sperb, 1980a).

Theorem 9.1 Let u be a positive $C^4(\Omega \times (0, T) \cap C^2(\bar{\Omega} \times [0, T))$ solution of (9.6) with $f \in C^1(\mathbb{R}^+), g, h \in C^2(\mathbb{R}^+)$. Suppose that

(i) $g(h/g)'' \geq 0, (h/g)(gf)' - h'f \geq 0$

for positive argument.

(ii)$_a$ If $g(1/g)'' \leq 0$, $(gf)'(1/g) \leq 0$ for positive argument, let $M = \max(M_0, M_1)$, where

$$\begin{aligned} M_0 &= \max_{\Omega} \{(\Delta u_0 + f(u_0))g(u_0) + h(u_0)\} \geq 0, \\ M_1 &= h(0) \geq 0. \end{aligned}$$

(ii)$_b$ Assume $M = 0$ if the conditions in (ii)$_a$ for g are not satisfied.

Then

$$g(u)u_{,t} + h(u) \leq M \quad \text{in} \quad \Omega \times (0, T). \tag{9.7}$$

Proof We show that $v(x,t) = g(u)u_{,t} + h(u)$ satisfies a parabolic maximum principle. We compute

$$\begin{aligned}\Delta v &= \Delta u_{,t} g + u_{,t}(g' \Delta u + g''|\nabla u|^2) + h' \Delta u + h''|\nabla u|^2 + 2g' \nabla u \cdot \nabla u_{,t},\\ v_{,t} &= \Delta u_{,t} g + f' g u_{,t} + g' u_{,t}^2 + h' u_{,t},\\ \nabla v &= g \nabla u_{,t} + u_{,t} g' \nabla u + h' \nabla u.\end{aligned}$$

After some manipulation we see that v satisfies

$$\begin{aligned}\Delta v - 2(\log g)' \nabla u \cdot \nabla v + \left[g \left(\frac{1}{g}\right)'' |\nabla u|^2 + (gf)' \frac{1}{g} \right] v - v_{,t} \\ = |\nabla u|^2 g \left(\frac{h}{g}\right)'' + \frac{h}{g}(gf)' - h'f. \qquad (9.8)\end{aligned}$$

Our assumption (i) guarantees that the right side in (9.8) is nonnegative, and thus the maximum principle for parabolic problems is applicable. If the inequality for $g(u)$, required in assumption (ii)_a, is satisfied, the maximum principle tells us that v cannot attain a nonnegative maximum M in the interior of $\Omega \times (0, T)$. If the condition for g in (ii)_a is not satisfied, we must have $M = 0$. Note that since $u(x,t) = 0$ on $\partial\Omega \times (0, T)$, we have $u_{,t} = 0$ on $\partial\Omega \times (0, T)$ for sufficiently smooth solutions (in particular, $u_0(x) = 0$ must hold for $x \in \partial\Omega$), and therefore $v(x,t) = h(0)$ for $x \in \partial\Omega$. The conclusion of Theorem 9.1 is then obvious.

Remarks on Theorem 9.1 (a) Theorem 9.1 can immediately be extended to solutions of

$$u_{,t} = Lu + f(u) \quad \text{in} \quad \Omega \times (0, T), \qquad (9.9)$$

where $Lu = a^{ij}(x)u_{,ij} + b^i(x)u_{,i}$ is a uniformly elliptic operator, under the same assumptions on $g(u)$, $h(u)$, and where in the assumption on u_0, Δ is replaced by L.

(b) The strong smoothness requirements for $u(x,t)$ can be relaxed if we proceed as in Sattinger (1973) and set

$$w_\delta(x,t) = g(u)\frac{u(x, t + \delta) - u(x,t)}{\delta} + h(u), \qquad \delta > 0.$$

The same calculation then leads to the inequality

$$\begin{aligned}\Delta w_\delta - 2(\log g)' \nabla u \nabla w_\delta + \left[g \left(\frac{1}{g}\right)'' |\nabla u|^2 + (g'f + gf'_\delta) \frac{1}{g} \right] w_\delta - w_{\delta,t} \\ = |\nabla u|^2 g \left(\frac{h}{g}\right)'' + \frac{h}{g}(g'f - gf'_\delta) - h'f, \qquad (9.10)\end{aligned}$$

which is the same as (9.8), with the only difference being that instead of $f'(u)$ we have

$$f'_\delta := \int_0^1 f_u(\tau u(x, t + \delta) + (1 - \tau)u(x, t))\, d\tau.$$

For more details refer to Sperb (1980a).

There are a number of simple choices of functions g, h that can be made depending on the properties of the nonlinearity $f(u)$. We shall illustrate this in a number of corollaries.

Corollary 9.1 (Sattinger, 1973) Let u be a positive solution of (9.6) and assume that $f \in C^1(\mathbb{R}^+)$ and

$$\Delta u_0 + f(u_0) \geq 0 \qquad \text{in} \quad \Omega.$$

Then

$$u_{,t} \geq 0 \qquad \text{in} \quad \Omega \times (0, T).$$

Proof With the simple choice $g(u) \equiv 1$, $h(u) \equiv 0$ and the assumption above on u_0, all requirements of Theorem 9.1 with $M = 0$ are satisfied. This proves Corollary 9.1.

Corollary 9.1 thus tells us that if $u(x, t)$ is initially increasing (decreasing) everywhere in Ω, it will remain so whenever a sufficiently smooth solution exists.

Remarks Corollary 9.1 remains true for other boundary conditions, e.g.,

$$\frac{\partial u}{\partial n} = 0 \qquad \text{or} \qquad \frac{\partial u}{\partial n} + \sigma(u) = 0, \qquad \sigma, \sigma' \geq 0.$$

The next result relates solutions of (9.1) to solutions of (9.3) under a somewhat restrictive condition on u_0. On the other hand, the resulting inequality is valid for every point in $\Omega \times (0, T)$ whenever a smooth solution exists.

Corollary 9.2 Let u be a sufficiently smooth positive solution of (9.6) and suppose $f \in C^2(\mathbb{R}^+)$, and

(i) $f(0) = 0$, $f(s) \geq 0$, and $f''(s) \geq 0$ for $s \geq 0$.
(ii) For some $a > 0$ we have

$$\Delta u_0 + f(u_0) \geq af(u_0) \qquad \text{in} \quad \Omega.$$

Then

$$u_{,t} \geq af(u) \qquad \text{in} \quad \Omega \times (0, T). \tag{9.11}$$

Proof With $g(u) \equiv 1$, $h(u) = -af(u)$, and the assumptions of Corollary 9.2, we can again apply Theorem 9.1, which implies inequality (9.11).

For the validity of inequality (9.11) f must be an increasing, convex function of its argument. It is possible to get inequalities similar to (9.11) if f is concave and positive. More precisely, we have

Corollary 9.3 Let u be a sufficiently smooth positive solution of (9.6), and suppose that

(i) $f(0) > 0, f(s) \geq 0, f''(s) \leq 0$ for $s \geq 0$.

(ii) $$M = \max_{\Omega} \frac{\Delta u_0 + f(u_0)}{f(u_0)} > 0,$$

$$m = \min_{\Omega} \frac{\Delta u_0 + f(u_0)}{f(u_0)} < 0.$$

Then

$$mf(u) \leq u_t \leq Mf(u). \tag{9.12}$$

Proof We choose $g(u) = 1/f(u)$, $h(u) \equiv 0$. Then the coefficient of v in Eq. (9.8) is nonpositive, provided that $f'' \leq 0$. On the other hand, the right-hand side in (9.8) becomes zero and thus allows us to apply the maximum principle if $M > 0$ and the minimum principle if $m < 0$. This proves the validity of (9.12).

As a final consequence of Theorem 9.1, we choose $g(u) \equiv 0$ and select $h(u)$ appropriately. It is not hard to check that for the special choice $g(u) \equiv 0$, i.e., $v(x,t) = h(u(x,t))$, the equation corresponding to (9.8) becomes

$$\Delta v - v_{,t} = h''|\nabla u|^2 - h'f. \tag{9.13}$$

This leads to the following result: Let $w(x,t)$ be the solution of the linear problem

$$\begin{aligned} w_{,t} &= \Delta w + 1 \quad &&\text{in} \quad \Omega \times (0,\infty), \\ w &= 0 \quad &&\text{on} \quad \partial\Omega, \\ w(x,0) &= w_0(x). \end{aligned} \tag{9.14}$$

Then we have

Corollary 9.4 Assume that u is a positive smooth solution of (9.6) and that

(i) $f \in C^1(\mathbb{R}^+)$ and $f' \geq 0$ for $s \geq 0$, $f(0) > 0$.

(ii) $$w(x,0) = \int_0^{u_0(x)} \frac{dy}{f(y)}.$$

Then

$$v(x,t) := \int_0^{u(x,t)} \frac{dy}{f(y)} \geq w(x,t). \tag{9.15}$$

Proof With $v(x,t) := \int_0^{u(x,t)} dy/f(y)$ we see that if $f' \geq 0$, then v satisfies

$$\Delta v + 1 - v_{,t} \leq 0 \qquad \text{in} \quad \Omega \times (0, T), \tag{9.16}$$

and v vanishes on $\partial\Omega$ since u vanishes there. Hence $v(x,t)$ cannot be smaller than the solution of (9.14), with the appropriate initial data. This proves Corollary 9.4.

Remark Note that the with solution $z(t)$ of (9.3) we have

$$v(t) := \int_{z_0}^{z(t)} \frac{dy}{f(y)} \qquad \text{and} \qquad v_{,t} = \dot{v} = 1 \qquad \text{by} \quad (9.4).$$

So far we have considered only Dirichlet boundary conditions. It is obvious that with some modifications we can prove results similar to Theorem 9.1 in the case in which u satisfies different boundary conditions.

Only minor changes are necessary in the proof of Theorem 9.1 to establish

Theorem 9.2 Let u be a sufficiently smooth solution of (9.6) where the boundary condition in (9.6) is replaced by $\partial u/\partial n = 0$ on $\Omega \times (0, T)$. Suppose that assumption (i) of Theorem 9.1 holds and that $(\text{ii})_\text{a}$, with $M = M_0$, or $(\text{ii})_\text{b}$ also holds. Then inequality (9.7) still holds.

Proof On $\partial\Omega$ we have for smooth solutions

$$\frac{\partial v}{\partial n} = u_{,t} g' \frac{\partial u}{\partial n} + g\left(\frac{\partial u}{\partial n}\right)_{,t} + h' \frac{\partial u}{\partial n} = 0,$$

and hence v must assume its maximum for $t = 0$ by the strong maximum principle. The remainder of the proof is the same as for Theorem 9.1.

Remarks (a) Remarks (a), (b) in Theorem 9.1 clearly apply here also.
(b) Corollary 9.1 holds again.
(c) Corollary 9.2 still holds if the Dirichlet boundary condition in (9.6) is replaced by the Neumann boundary condition. At the same time $f(0) = 0$ is not necessary in this case.
(d) Corollaries 9.3 and 9.4 also carry over with some slight and obvious modifications.

Next we assume that the boundary condition in (9.6) is replaced by

$$\frac{\partial u}{\partial n} + \sigma(u) = 0 \qquad \text{on} \quad \partial\Omega \times (0, T). \tag{9.17}$$

Then we have

Theorem 9.3 Let u be a sufficiently smooth, positive solution of (9.6) with the boundary condition (9.17) instead of Dirichlet boundary conditions.

Suppose that assumption (i) of Theorem 9.1 holds and $(\text{ii})_a$, $(\text{ii})_b$, with $M = M_0$, as well as (iii). For positive arguments we have $g, \sigma > 0$ and

$$(\log \sigma)' + (\log g)' \geq 0, \qquad (\sigma h)' + \sigma h(\log g)' \leq 0.$$

Then

$$g(u)u_{,t} + h(u) \leq M \qquad \text{in} \quad \Omega \times (0, T).$$

Proof It suffices to ensure that v cannot attain its maximum on $\partial\Omega$. On $\partial\Omega$ we find for smooth solutions u

$$\frac{\partial v}{\partial n} = g' \frac{\partial u}{\partial n} u_{,t} + \frac{\partial (u_{,t})}{\partial n} g + h' \frac{\partial u}{\partial n}.$$

But by (9.17) we have

$$\frac{\partial (u_{,t})}{\partial n} + \sigma' u_{,t} = 0 \qquad \text{on} \quad \partial\Omega,$$

and hence

$$\frac{\partial v}{\partial n} = -\sigma g' u_{,t} - \sigma' u_{,t} g - h'\sigma = -\sigma g' \frac{v - h}{g} - \sigma'(v - h) - h'\sigma,$$

so that we find

$$\frac{\partial v}{\partial n} + (\sigma' + \sigma(\log g)')v = (\sigma h)' + \sigma h(\log g)'.$$

But if $\sigma' + \sigma(\log g)' \geq 0$ and $(\sigma h)' + \sigma h(\log g)' \leq 0$, it follows from the strong maximum principle that v cannot assume its maximum on $\partial\Omega$. Therefore, it must be attained for $t = 0$. This proves Theorem 9.3.

It is quite obvious how Corollaries 9.2–9.4 have to be modified in the case of boundary condition (9.17). As an example we state the analog of Corollary 9.2 as

Corollary 9.5 Let u be a sufficiently smooth positive solution of (9.6) satisfying (9.17) instead of $u = 0$ on $\partial\Omega$. If the nonlinearity $f \in C^1(\mathbb{R}^+)$ satisfies

(i) $f'' \geq 0$ for positive argument,
(ii) $\sigma' \geq 0$, $(\sigma f)' \geq 0$ for positive argument,
(iii) $\Delta u_0 + f(u_0) \geq af(u_0)$ for some $a > 0$,

then

$$u_{,t} \geq af(u) \qquad \text{in} \quad \Omega \times (0, T).$$

Proof As in Corollary 9.2, we have chosen $g(u) \equiv 1$, $h(u) = -af(u)$. Then it is easy to check that Theorem 9.3 is valid for $v = u_{,t} - af(u)$ if the assumptions (i)–(iii) are satisfied.

Remarks on Theorems 9.1–9.3 (a) The inequalities in Theorems 9.1–9.3 are optimal in the following sense: If ϕ_1 is the first (nonconstant) eigenfunction of

$$\Delta\phi + \lambda\phi = 0 \qquad \text{in} \quad \Omega,$$

satisfying one of the boundary conditions of Theorems 9.1–9.3 (with $\sigma(u) = \alpha u$ and α a positive constant in (9.17)), $u_0(x) = \phi_1(x)$, and $f(u) = \gamma u$, $\gamma > 0$, then the solution of (9.6) is given by

$$u(x, t) = e^{(\gamma - \lambda_1)t}\phi_1(x)$$

and

$$u_{,t} = (\gamma - \lambda_1)f(u),$$

i.e., the equality sign holds in Corollary 9.2 (and its analogs) with $a = (\gamma - \lambda_1)/\gamma$.

(b) One could modify the assumptions of Theorems 9.1–9.3 so that functions $f(u, x)$ would be allowed. The details are left to the interested reader.

(c) It is not hard to extend the results to the case in which, e.g., u is a solution of

$$u_{,t} = D(t)\Delta u + f(u),$$

i.e., to introduce a time-dependent diffusion coefficient (see Sperb (1979b)). One could also handle equations of the form

$$u_{,t} = \operatorname{div}(D(u)\nabla u) + f(u)$$

in a similar way.

9.2 APPLICATIONS

9.2.1 Bounds for the "Blow-Up Time"

The simple example of the equation

$$\dot{z} = z^2, \qquad t > 0,$$
$$z(0) = 1,$$

whose solution is given by

$$z(t) = \frac{1}{1 - t},$$

displays already an important feature that is common to a large class of reaction-diffusion equations, i.e., the solution becomes unbounded for a finite time.

As an application of the results of Section 9.1, we shall give bounds for the blow-up time T. A first result can be stated as

Corollary 9.6 Assume that $f(s) \in C^2$ for $s > 0$ and that

(i) $f(0) = 0, f(s) > 0, f''(s) > 0$ for $s > 0$,
(ii) $0 \leq u_0(x) \leq M_0$ for $x \in \Omega$ and $\Delta u_0 + f(u_0) \geq af(u_0)$ in Ω, $a \in (0, 1)$
(iii) $\displaystyle\int_{M_0}^{\infty} \frac{dy}{f(y)} < \infty$.

Then any positive $C^4(\Omega \times (0, T)) \cap C^2(\bar{\Omega} \times [0, T))$ solution of (9.6) must blow up and

$$T \leq \frac{1}{a} \int_{M_0}^{\infty} \frac{dy}{f(y)}.$$

Proof Since the assumptions of Corollary 9.2 are satisfied, inequality (9.11) holds for every point (x, t) in $\Omega \times (0, T)$, i.e.,

$$u_{,t} \geq af(u) \qquad \text{in} \quad \Omega \times (0, T).$$

In particular, at the point x_0 at which $u_0(x_0) = M_0$ we get by integration

$$\int_{M_0}^{u(x_0, t)} \frac{dy}{f(y)} \geq at.$$

But if assumption (iii) of Corollary 9.6 is satisfied, we would be led to a contradiction for any time

$$t > \frac{1}{a} \int_{M_0}^{\infty} \frac{dy}{f(y)},$$

which proves Corollary 9.6.

Remarks It is not hard to check that the inequality

$$\int_{M_0}^{u(x_0,t)} \frac{dy}{f(y)} \geq at$$

gives a lower bound for $u(x_0, t)$ that tends to infinity as t tends toward the value

$$\frac{1}{a} \int_{M_0}^{\infty} \frac{dy}{f(y)}.$$

This shows that any sufficiently smooth solution of (9.6) will eventually become unbounded for a finite value of t. Of course the possibility that a solution would cease to be as smooth as is required in Corollary 9.6 before the blow-up time is reached is not excluded. The interested reader is referred to the paper of Ball (1977) where this problem is discussed.

Examples (a) Assume that in addition to hypothesis (i) of Corollary 9.6, we have $f'(0) > \lambda_1$, where λ_1 is the first eigenvalue of $\Delta\phi + \lambda\phi = 0$ in Ω, $\phi = 0$ on $\partial\Omega$, and ϕ_1 is the corresponding first eigenfunction with $\max_\Omega \phi(x) = \phi_M$. For $u_0(x) = \phi_1(x)$ it is easy to check that we may take

$$a = 1 - \frac{\lambda_1}{f'(0)},$$

and hence

$$T \le \frac{f'(0)}{f'(0) - \lambda_1} \int_{\phi_M}^{\infty} \frac{dy}{f(y)}.$$

In particular, for $f(s) = \lambda(s + s^p)$, $p > 1$, $\lambda > \lambda_1$, we find

$$T \le \frac{\log 2}{(\lambda - \lambda_1)p}.$$

(b) It follows from the results of Levinson (1962) that the problem

$$\Delta\psi + \mu\psi^p = 0 \quad \text{in} \quad \Omega, \qquad \psi = 0 \quad \text{on} \quad \partial\Omega, \tag{9.18}$$

has a positive solution for any $\mu > 0$ if $p > 1$.

Assume now that we are concerned with problem (9.6) and $f(s) = \lambda s^p$. Suppose $u_0(x)$ is a steady state of (9.6), i.e., a solution of (9.18) for some value $\lambda = \mu_0$, and λ has increased from $\lambda = \mu_0$ to a larger value $\lambda = \mu_1$. Then it is easy to check that now $a = 1 - \mu_0/\mu_1$ and $u(x, t)$ will blow up not later than

$$T^* = \frac{p - 1}{(\mu_1 - \mu_0)\psi_M}, \qquad \psi_M = \max_\Omega \psi(x).$$

At this point we can make use of inequality (6.63), which gives

$$\psi_M \ge \left\{\rho\left(\frac{2\mu_0}{p + 1}\right)^{1/2} N(p)\right\}^{-2/(p-1)},$$

with

$$N(p) = \frac{1}{\sqrt{\pi}}\left(\left(\Gamma\left(\frac{1}{p + 1}\right)\right)^{-1} (p + 1)\Gamma\left(\frac{p + 3}{2(p + 1)}\right)\right),$$

and ρ the radius of the largest circle inscribed in Ω.

Corollary 9.4 may also be used in order to prove a nonexistence result in the following way. This result may be stated as

Corollary 9.7 Let u be a positive smooth solution of (9.6) and suppose that

(i) $f(0) > 0$ and $f' \geq 0$ for positive argument.
(ii) $\int_0^\infty dy/f(y) < \psi_M$, $\psi_M = \max_\Omega \psi$ and ψ is the solution of $\Delta\psi + 1 = 0$ in Ω, $\psi = 0$ on $\partial\Omega$.

Then no smooth solution (9.6) can exist for all time.

Proof By Corollary 9.4 we know that

$$\int_0^{u(x,t)} \frac{dy}{f(y)} \geq w(x, t).$$

but for any finite initial data $u_0(x)$, $w(x, t)$ tends to $\psi(x)$ as $t \to \infty$. The inequality in assumption (ii) would then lead to a contradiction if a solution is supposed to exist for all time.

Remarks (a) To get a more explicit result one may use any of the lower bounds derived for ψ_M.
(b) Note that no estimate for the existence time is given. On the other hand, however, any sufficiently smooth initial data are allowable.

Let us now consider an absorption process modeled by

$$\begin{aligned} u_{,t} &= \Delta u && \text{in} && \Omega \times (0, T), \\ \frac{\partial u}{\partial n} &= \sigma(u) && \text{on} && \partial\Omega \times (0, T), \\ u &= u_0 && && t = 0. \end{aligned} \tag{9.19}$$

Again, one may expect a blow-up of the solution under certain circumstances. Another simple choice of the functions $g(u)$, $h(u)$ in Theorem 9.3 then yields

Corollary 9.8 Suppose that u is a smooth positive solution of (9.19) and

(i) $\sigma, \sigma''(s) \geq 0$ for $s > 0$,
(ii) $0 \leq u_0 \leq M_0$ and for some $a > 0$, $M_0 \geq 0$ $\Delta u_0 \geq a\sigma(u_0)$ in Ω.
(iii) $\displaystyle\int_{M_0}^\infty \frac{dy}{\sigma(y)} < \infty.$

Then $u(x, t)$ must become infinite for some finite time T, and

$$T \leq \frac{1}{a} \int_{M_0}^\infty \frac{dy}{\sigma(y)}.$$

Proof We choose $g = -1/\sigma$ and $h = a = \text{const}$ in Theorem 9.3. Under hypotheses (i) and (ii), the statement of Theorem 9.3 holds with $M = 0$. Hypothesis (ii) and a simple integration immediately imply the result stated in Corollary 9.8.

Example Take $\sigma(u) = \gamma u^2$, $\gamma > 0$, and $u_0 = r^2$, where r measures the distance from the origin. Suppose Ω is some domain in E^N. Then the hypotheses of Corollary 9.8 are satisfied with $a = 2N/\gamma R^2$, where $R = \max_\Omega r$, and $T \leq 1/2N$. Note that in (9.19) t is a dimensionless "time" and u has the dimension square centimeters.

9.2.2 Pointwise Bounds

The results of Section 9.1 can also be used in order to obtain pointwise bounds for the solution of (9.6).

Example 1 Let u be a solution of (9.6) and

$$f(s) = v\left(s + \frac{bs}{1+s}\right).$$

Take $u_0(x)$ as the first eigenfunction $\phi_1(x)$ of

$$\Delta\phi + \lambda\phi = 0 \quad \text{in} \quad \Omega, \qquad \Phi = 0 \quad \text{on} \quad \partial\Omega.$$

Corollary 9.2 is applicable with $a = 1 - \lambda_1/(f'(0)) = 1 - \lambda_1/(1+b)$, and it tells us that for any $x \in \Omega$ we have

$$\int_{u_0(x)}^{u(x,t)} \frac{dy}{f(y)} \geq at, \qquad \text{if} \quad f'' \geq 0$$

and

$$\int_{u_0(x)}^{u(x,t)} \frac{dy}{f(y)} \leq at, \qquad \text{if} \quad f'' \leq 0.$$

An elementary integration then leads to the inequality

$$u(x,t)(u(x,t) + 1 + b)^b \geq \phi_1(x)(\phi_1(x) + 1 + b)^b \times \{(v(1+b) - \lambda_1)t\}$$

if $b \geq 0$ and $v(1+b) > \lambda_1$, but

$$u(x,t)(u(x,t) + 1 + b)^b \leq \phi_1(x)(\phi_1(x) + 1 + b)^b \times \{(v(1+b) - \lambda_1)t\}$$

if $-1 < b < 0$. Both inequalities become equalities for $b = 0$.

Example 2 We consider a solution of (9.19) and apply Theorem 9.3, choosing $g(u) = -1/\sigma(u)$, $h(u) = a\sigma'(u)$, and $M = 0$. One easily checks that the assumptions of Theorem 9.3 are satisfied, provided that

(i) $\sigma \geq 0$, $\sigma' \geq 0$, $\sigma'' \leq 0$, $(\sigma^2)''' \geq 0$ for positive argument,
(ii) $\Delta u_0 \geq a\sigma(u_0)\sigma'(u_0)$ in Ω for some $a > 0$.

Then for the solution $u(x, t)$ of (9.19) one has

$$\int_{u_0(x)}^{u(x,t)} \frac{dy}{\sigma(y)\sigma'(y)} \geq at.$$

On the other hand, if $\sigma \geq 0$, $\sigma' \geq 0$, $a > 0$, with all the other inequality signs reversed in assumptions (i), (ii), then

$$\int_{u_0(x)}^{u(x,t)} \frac{dy}{\sigma(y)\sigma'(y)} \leq at$$

holds. Note that the bounds for $u(x, t)$ given in the implicit form above remain finite for any finite time, provided that the integral

$$\int_0^\infty \frac{dy}{\sigma(y)\sigma'(y)}$$

diverges. It is interesting to note that if this latter integral converges, the solution $u(x, t)$ may blow up for finite time, as was shown by Walter (1975). It is not hard to check that if σ, $\sigma' \geq 0$, $\sigma'' \leq 0$, then the integral $\int^\infty dy/\sigma(y)\sigma'(y)$ cannot converge. However, it remains, of course, finite if σ, $\sigma' \geq 0$, but $\sigma'' > 0$. Then the blow-up time T is bounded below by

$$\frac{1}{a}\int_{m_0}^\infty \frac{dy}{\sigma(y)\sigma'(y)},$$

where $m_0 = \min_\Omega u_0$.

9.3 MAXIMUM PRINCIPLES FOR $P = |\nabla u|^2 + 2z(t)F(u)$

Of course it is harder to get estimates for the gradient for solutions of nonlinear parabolic problems. The choice of our function P made in this section is just a slight modification of the simplest one used in the corresponding steady state. One could choose more general forms such as, e.g.,

$$P = g(u)|\nabla u|^2 + z(t)h(u),$$

but for the sake of simplicity we restrict our attention to the simpler version $P = |\nabla u|^2 + 2z(t)F(u)$.

Similar to the elliptic case, we first prove

Lemma 9.1 Let u be a positive $C^3(\Omega \times (0, T) \cap C^2(\bar{\Omega} \times [0, T))$ solution of (9.6), and suppose that the following conditions hold:

(i) $\partial\Omega$ has nonnegative mean curvature H.

(ii) $f(0) = 0$, $f' \geq 0$, $(4(F^{1/2})'' + \alpha F^{1/2})(z_0 - 1) \leq 0$ for some finite α and some $z_0 > 0$ $(F' = f)$.

(iii) $z(t) = \{1 - (1 - z_0{}^{-1})e^{\alpha t}\}^{-1}$.

Then the function

$$P(x, t) = |\nabla u|^2 + 2z(t)F(u)$$

assumes its maximum in $\Omega \times (0, T)$ for $t = 0$ or at a point where $\nabla u = 0$.

Proof By differentiation we get

$$P_{,k} = 2u_{,i}u_{,ik} + 2zfu_{,k}, \tag{9.20}$$

$$\Delta P = 2u_{,ik}u_{,ik} + 2zf'|\nabla u|^2 + 2zf\,\Delta u + 2u_{,i}(\Delta u)_{,i}, \tag{9.21}$$

$$P_{,t} = 2u_{,i}(\Delta u + f)_{,i} + 2\frac{dz}{dt}F + 2zf(\Delta u + f). \tag{9.22}$$

Schwarz's inequality and (9.20) give

$$4u_{,i}u_{,i}u_{,jk}u_{,jk} \geq (P_{,k} - 2zfu_{,k})(P_{,k} - 2zfu_{,k}), \tag{9.23}$$

so that after some calculation we see that

$$\Delta P - \frac{1}{|\nabla u|^2}\left\{\frac{1}{2}\nabla P - 2zf\,\nabla u\right\} \cdot \nabla P - 2(z-1)f'P - P_{,t}$$
$$\geq 2F\left\{z(z-1)\left(\frac{f^2}{F} - 2f'\right) - \frac{dz}{dt}\right\} = -2F\left\{z(z-1)4(F^{1/2})''F^{-1/2} + \frac{dz}{dt}\right\}. \tag{9.24}$$

The function $z(t)$ as chosen in assumption (iii) satisfies

$$\frac{dz}{dt} = \alpha z(z-1), \qquad z(0) = z_0.$$

Clearly, if $0 < z_0 < 1$, then $z(t)$ lies in the same interval, and the same is true if $z_0 > 1$. If one assumes as in hypothesis (ii) that

$$(4(F^{1/2})'' + \alpha F^{1/2})(z_0 - 1) \leq 0,$$

then P satisfies a parabolic inequality of the form

$$\Delta P - \frac{L_i P_{,i}}{|\nabla u|^2} - hP - P_{,t} \geq 0 \qquad \text{in} \quad \Omega \times (0, T), \tag{9.25}$$

with the obvious definitions of the terms L_i and h. If $(z(t) - 1)f' \geq 0$, we can apply the maximum principle, which leads to the following possible cases where P may assume its maximum:

(a) for $t = 0$,
(b) at a point where $\nabla u = 0$,
(c) on the boundary $\partial\Omega$.

It remains to show that case (c) can be excluded if the average curvature H of $\partial\Omega$ is nonnegative.

Now if $u = 0$ on $\partial\Omega \times (0, T)$, then $u_{,t} = 0$ on $\partial\Omega \times (0, T)$ for smooth solutions, and we find with the same calculations as in the elliptic case that

$$\frac{\partial P}{\partial n} = 2\frac{\partial u}{\partial n}\frac{\partial^2 u}{\partial n^2} + 2zf(0)\frac{\partial u}{\partial n} = -2\left(\frac{\partial u}{\partial n}\right)^2 (N-1)H \leq 0 \tag{9.26}$$

because of our assumptions that $H \geq 0$ and $f(0) = 0$. This proves Lemma 9.1.

Remarks (a) If we choose $z_0 = 1$, i.e., $z(t) \equiv 1$, then assumptions (ii) and (iii) can be dropped.

(b) It follows from equation (9.26) that the statement of Lemma 9.1 remains true if $H \geq H_0 > 0$, $f(0) > 0$, and $\tau(N-1)H_0 + (z(t) - 1)f(0) > 0$, where $\tau = \max_{\partial\Omega} |\nabla u|$. The remaining assumptions in Lemma 9.1 are of course kept.

In Lemma 9.1 we still do not know exactly where the function P assumes its maximum. Let us now look at the case in which P assumes its maximum at a point $x \in \Omega$, where $\nabla u = 0$ for some $t > 0$. At this point x we have

$$P_{,t} = 2\left(\frac{dz}{dt}F + zfu_{,t}\right) = 2zF\left(\alpha(z-1) + \frac{f}{F}u_{,t}\right). \tag{9.27}$$

It is clear then that if we can give sufficient conditions ensuring that $P_{,t} \leq 0$ for $t > 0$, it will follow that P must assume its maximum initially. On the other hand, if we know that P cannot attain its maximum for $t = 0$, and $P_{,t} \geq 0$ for $t > 0$, it follows that the maximum of P in any subdomain $\Omega \times (0, t_1)$ of $\Omega \times (0, T)$ $(t_1 < T)$ must occur at a critical point of u at the "latest time" $t = t_1$.

A very simple case is considered in the next result.

Theorem 9.4 Let u be a positive, sufficiently smooth solution of (9.6), and suppose that the following hypotheses hold:

(i) $\partial\Omega$ has nonnegative mean curvature.
(ii) The initial function u_0 satisfies $\Delta u_0 + f(u_0) \leq 0$ in Ω.

Then the function

$$P = |\nabla u|^2 + 2F(u)$$

attains its maximum for $t = 0$.

Proof We simply combine Corollary 9.1, Lemma 9.1 (see Remark (a) on Lemma 9.1), and (9.27), which shows that $P_{,t} \leq 0$ would hold for any $t > 0$ at a critical point of u. This proves Theorem 9.4.

Next we construct a case in which the maximum is attained at the "latest time." First, the following auxiliary result is needed.

Lemma 9.2 Let w be any $C^3(\Omega) \cap C^2(\bar{\Omega})$ function satisfying

$$\Delta w + g(w) \geq 0 \qquad \text{in} \quad \Omega,$$

and also

$$\nabla w \cdot \nabla(\Delta w + g(w)) \geq 0 \qquad \text{in} \quad \Omega,$$

and $w = 0$ on $\partial\Omega$. Here g is an arbitrary nonnegative C^1 function of its argument. Suppose that the mean curvature of $\partial\Omega$ is nonnegative. Then, the function

$$P = |\nabla w|^2 + 2G(w) \qquad (G' = g)$$

attains its maximum at a point where $\nabla w = 0$.

Proof Calculations similar to those used in Chapter 5 show that P satisfies

$$\Delta P - \frac{1}{2|\nabla w|^2} \nabla P \cdot (\nabla P - 4g\nabla w) \geq 2g(w)(\Delta w + g(w)) + 2\nabla w \cdot \nabla(\Delta w + g(w)) \geq 0$$

by assumption. On the boundary we find as before that

$$\frac{\partial P}{\partial n} \leq 0,$$

which proves Lemma 9.2.

We can now prove

Theorem 9.5 Let u be a sufficiently smooth positive solution of (9.6), and suppose that the following conditions are satisfied.

(i) $\partial\Omega$ has nonnegative mean curvature.

(ii) $f(0) = F(0) = 0$, $f'(0) > 0$, and for positive argument and some $\alpha > 0$ we have $f'' \geq 0$ and $0 \leq 4F^{-1/2}(F^{1/2})'' \leq \alpha$.

(iii) The initial data u_0 satisfies $u_0 = 0$ on $\partial\Omega$ and $\Delta u_0 + f(u_0) \geq af(u_0)$, $\nabla u_0 \cdot \nabla(\Delta u_0 + f(u_0)) \geq 0$ in Ω with $a > 0$.

(iv) $z(t) = ((1 - (1 - z_0^{-1})e^{-\alpha t})^{-1}$, $z_0 > 1$, and $\alpha(z_0 - 1) \leq 2f'(0)a$.

Then the function

$$P = |\nabla u|^2 + 2z(t)F(u)$$

attains its maximum in $\Omega \times (0, t_1)$, $t_1 < T$ at a critical point of u for $t = t_1$.

Proof We make use of Corollary 9.2 first, which gives the estimate

$$u_{,t} \geq af(u),$$

so that from equation (9.27) it follows that (replace α by $-\alpha$!)

$$P_{,t} \geq 2zF\left(-\alpha(z - 1) + a\frac{f^2}{F}\right).$$

But

$$\left(\frac{f^2}{F}\right)' = \frac{f}{F}\left(2f' - \frac{f^2}{F}\right) = \frac{f}{F}4(F^{1/2})''F^{-1/2} \geq 0$$

and

$$\lim_{s\to 0}\frac{f^2(s)}{F(s)} = 2f'(0) > 0.$$

Furthermore, since $z(t) > 1$ if $z_0 > 1$, and $z(t)$ as chosen in assumption (iv) is decreasing for t increasing, it follows that $P_{,t} \geq 0$ for any $t > 0$, provided that

$$2af'(0) \geq \alpha(z_0 - 1)$$

holds.

Finally, the maximum of P for $t = 0$ also occurs at a point where $\nabla u_0 = 0$ because of assumption (ii) and Lemma 9.2. This completes the proof.

Remarks (a) One easily checks that Theorem 9.5 remains true if instead of (ii) one has

(ii′) $f(0) = 0$, $f \geq 0$ for positive argument,

and (iv) is dropped. Then the function

$$P = |\nabla u|^2 + 2F(u)$$

attains its maximum in $\Omega \times (0, t_1)$, $t_1 < T$, at a critical point of u for $t = t_1$.

(b) If the assumptions of Theorem 9.5 are satisfied, then we get in particular

$$z_0F(m_0) \leq z(t)F(u_m(t)), \tag{9.28}$$

where

$$m_0 = \max_{\Omega} u_0, \qquad u_m(t) = \max_{\Omega} u(x,t).$$

On the other hand, by Remark (a) we know that

$$F(m_0) \le F(u_m(t)), \tag{9.29}$$

which is cruder than (9.28) since $z(t)$ as chosen in Theorem 9.5 is decreasing for increasing t.

Next we consider a situation for which $z(t)$ is decreasing, $u(x,t)$ may be increasing in t at any point, but for which the function P still assumes its maximum initially. More precisely, we have

Theorem 9.6 Let u be a sufficiently smooth solution of (9.6), and suppose that

(i) The mean curvature H of $\partial\Omega$ satisfies $H \ge H_0 > 0$.
(ii) $f(0) > 0$, $F(0) > 0$, and for positive argument we have $f \ge 0$, $f' \le 0$, $f'' \le 0$ and $(f^2/F) - 2f' \le \alpha$, $f^2/F \le \beta$.
(iii) The initial data satisfy $u_0 = 0$ on $\partial\Omega$ and

$$\max_{\Omega} \frac{\Delta u_0 + f(u_0)}{f(u_0)} = M > 0.$$

(iv) $z(t) = \{1 - (1 - z_0^{-1})e^{\alpha t}\}^{-1}$, and

$$\frac{M\beta}{\alpha} \le 1 - z_0 \le \frac{\tau(N-1)}{f(0)} H_0.$$

Then the function

$$P = |\nabla u|^2 + 2z(t)F(u)$$

assumes its maximum at $t = 0$.

Proof We have only to check the possibility that P attains its maximum at a critical point of u for $t > 0$. Our assumptions (ii) and (iii) allow the application of Corollary 9.3, so that the following estimate is valid

$$u_{,t} \le Mf.$$

Hence by (9.27) it follows that

$$P_{,t} = 2zF\left(\alpha(z-1) + \frac{f}{F}u_{,t}\right) \le 2zF\left(\alpha(z-1) + M\frac{f^2}{F}\right)$$
$$\le 2zF(\alpha(z-1) + M\beta.$$

Since $z(t)$ is decreasing in our case, it suffices to have

$$\alpha(z_0 - 1) + M\beta \leq 0$$

to conclude that $P_{,t} \leq 0$, which proves Theorem 9.6.

Concluding Remarks (a) So far we have only considered Dirichlet boundary conditions when dealing with maximum principles for functions

$$P = |\nabla u|^2 + 2z(t) \cdot F(u).$$

It is quite clear that one can easily give analogs of our last three theorems for other types of boundary conditions. The only new feature then is that equation (9.26) will have to be changed.

A simple example is the following result:

Theorem 9.7 Let u be a sufficiently smooth solution of

$$u_{,t} = \Delta u + f(u) \qquad \text{in} \quad \Omega \times (0, T),$$

$$\frac{\partial u}{\partial n} = 0 \qquad \text{on} \quad \partial\Omega \times (0, T),$$

$$u(x, 0) = u_0(x)$$

where it is assumed that

(i) $\partial\Omega$ is convex, and
(ii) u_0 satisfies $\Delta u_0 + f(u_0) \leq 0$ in Ω.

Then, the function

$$P = |\nabla u|^2 + 2F(u)$$

assumes its maximum for $t = 0$.

Proof The same calculations as in Chapter 5 give (see equation (5.44))

$$\frac{\partial P}{\partial n} = -2 \sum_{p=1}^{N-1} \kappa_p \left(\frac{\partial u}{\partial s_p}\right)^2 \leq 0, \tag{9.30}$$

if $\partial\Omega$ is convex. At a critical point of u for $t > 0$ we have again $P_{,t} \leq 0$ because of Corollary 1. This proves Theorem 9.7.

(b) On the basis of Chapter 8, it is clear how the results can be extended to the case in which the Laplacian is replaced by a uniformly elliptic operator L. One will then consider functions

$$P = a^{ij} u_{,i} u_{,j} + 2z(t)F(u),$$

where the a^{ij} are the coefficients of the elliptic operator L. This straightforward extension is left to the interested reader.

(c) It would also be possible to find functions P, say of the form

$$P = |\nabla u|^2 + z(t)F(u),$$

such that their maximum is attained on $\partial\Omega \times (0, T)$. In order to obtain these results, one can again make use of the fact that

$$u_{,ik}u_{,ik} \geq \frac{1}{N}(\Delta u)^2$$

in the place of Schwarz's inequality in (9.21).

(d) In two space-dimensions one could also use the identity (5.15) in the place of Schwarz's inequality. Instead of the parabolic inequality (9.24), one would then get a parabolic equation.

(e) Further extensions may be interesting too. For example, if the differential equation is

$$u_{,t} = (v(q)u_{,i})_{,i} + w(q)f(u), \qquad q = |\nabla u|^2, \tag{9.31}$$

an obvious selection of a function P would be

$$P = \int_0^q \frac{2sv'(s) + v(s)}{w(s)}\,ds + 2z(t)\int_0^u f(y)\,dy. \tag{9.32}$$

9.4 APPLICATIONS: BOUNDS FOR $|\nabla u|$

The main purpose to derive the results of Section 9.3 is that they allow to derive bounds for the absolute value of the gradient. The following result is a corollary of Theorem 9.5.

Corollary 9.9 Let u be a sufficiently smooth positive solution of problem (9.6), and suppose that

(i) $f(0) = F(0) = 0$, $f'(0) > \lambda_1$, $f''(s) \leq 0$, $f'''(s) \leq 0$ for $s \geq 0$, and $\lim_{s\to\infty} f'(s) =: f'(\infty) > f'(0)$, $f'(\infty) < \infty$.

(ii) $u_0 = \phi_1$ with corresponding eigenvalue λ_1.

(iii) The mean curvature H of $\partial\Omega$ is nonnegative. Then, for any $t_1 > t > 0$ we have

$$|\nabla u(x,t)|^2 + 2z(t)F(u(x,t)) \leq 2z(t_1)F(u_m(t_1))$$

with

$$z(t) = (1 - (1 - z_0^{-1})e^{-\alpha t})^{-1}$$

with

$$\alpha = \max_{s>0}\left(2f'(s) - \frac{f^2(s)}{F(s)}\right), \qquad 0 < \alpha < \infty$$

and

$$z_0 = 1 + \frac{2}{\alpha}(f'(0) - \lambda_1).$$

Proof We only have to check that the assumptions of Theorem 9.5 are satisfied. To this end we note that for

$$\phi(s) := 2f'(s) - \frac{f^2(s)}{F(s)}$$

we have

$$\lim_{s \to 0} \Phi(s) = \lim_{s \to \infty} \Phi(s) = 0.$$

We show that $\Phi(s) > 0$ for $s > 0$.

Contrariwise, assume that $\Phi(s)$ attains a negative minimum for some $s^* > 0$. For this value of s we have

$$\begin{aligned} \left.\frac{d^2\Phi}{ds^2}\right|_{s^*} &= 2f'''(s^*) - \frac{1}{F(s^*)}\left(f'(s^*) - \frac{f^2(s^*)}{F(s^*)}\right)\Phi(s^*) \\ &\quad - \frac{f(s^*)}{F(s^*)}\frac{d\Phi}{ds}(s^*) \\ &= 2f'''(s^*) - \frac{1}{f(s^*)}\left(f'(s^*) - \frac{f^2(s^*)}{F(s^*)}\right)\Phi(s^*). \end{aligned}$$

But if $\Phi(s^*) < 0$ and $f'''(s) < 0$, it would clearly follow that $d\Phi/ds < 0$ at $s = s^*$, thus contradicting our assumption. So $\Phi(s)$ assumes a finite positive maximum for $s > 0$. Next we check assumption (ii) of Theorem 9.5. Since

$$\Delta\phi_1 + f(\phi_1) = f(\phi_1) - \lambda_1\phi_1 =: g(\phi_1)$$

and

$$g(0) = 0, \qquad g'(0) > 0, \qquad g''(s) \geq 0 \qquad \text{for} \quad s > 0,$$

we see that

$$\Delta\phi_1 + f(\phi_1) > af(\phi_1), \qquad \nabla\phi_1 \cdot \nabla(\Delta\phi_1 + f(\phi_1)) = g'(\phi_1)|\nabla\phi_1|^2 \geq 0$$

if we choose $a = 1 - \lambda_1/(f'(0))$. Finally, we take $z_0 = 1 + 2f'(0)(a/\alpha)$ with $\alpha = \max_{s>0} \Phi(s)$, and all hypotheses of Theorem 9.5 are then satisfied.

Example For $f(s) = \lambda(s - (0.5\, s)/(1 + s))$ we need $\lambda > 2\lambda_1$. A numerical approximation gives $\alpha = 0.12$, and

$$z_0 = 1 + 8.33\left(1 - 2\frac{\lambda_1}{\lambda}\right).$$

Corollary 9.9 gives the bound

$$|\nabla u|^2(x,t) \leq 2z(t)\{F(u_m(t) - F(u(x,t))\}$$

for any $t > 0$. In particular, we have

$$\tau^2(t) := \max_{\partial\Omega} |\nabla u(x,t)|^2 \leq 2z(t)F(u_m(t)),$$

where a bound for $u_m(t)$ can be calculated as in Section 9.2.2 giving

$$u_m(t) \leq \tfrac{1}{2}n(t)(n(t) + (n^2(t) + 2)^{1/2}),\ n(t) := (\tfrac{2}{3})^{1/2} \exp((\tfrac{1}{2}\lambda - \lambda_1)t)$$

and

$$F(s) = \tfrac{1}{2}\lambda(s^2 - s + \log(1 + s)).$$

Chapter 10
FURTHER EXTENSIONS

10.1 FOURTH-ORDER PROBLEMS

It is fairly obvious that we cannot expect immediate extensions to equations of higher order than two since in general there are no corresponding maximum principles available in that case. Still a number of similar results can be proven. Let us first look at a result by Miranda (1948).

Theorem 10.1 (Miranda) Let u be a sufficiently smooth function satisfying $\Delta^2 u = 0$ in a plane domain D. Then the function

$$P := |\nabla u|^2 - u\,\Delta u$$

attains its maximum on ∂D.

Proof We have

$$\begin{aligned} P_{,k} &= 2u_{,ik}u_{,i} - u_{,k}\,\Delta u - u\,\Delta u_{,k}, \\ \Delta P &= 2u_{,ikk}u_{,i} + 2u_{,ik}u_{,ik} - (\Delta u)^2 - 2u_{,k}(\Delta u)_{,k} - u\,\Delta^2 u \\ &= 2u_{,ik}u_{,ik} - (\Delta u)^2 \geq 0 \end{aligned}$$

by Lemma 5.4. Hence the maximum principle implies the statement of Theorem 10.1.

Remarks (a) If u is a solution of

$$\Delta u = \text{const} \qquad \text{in} \quad D,$$

then Theorem 10.1 is contained in Theorem 5.1.

(b) If we change P slightly into

$$P = |\nabla u|^2 - 2u\,\Delta u,$$

then it follows that P *cannot* assume its maximum on ∂D if the average curvature of $\partial D\,(D \subset E^N, N \geq 2)$ is positive and if u vanishes on ∂D (unless $u \equiv \text{const}$), as was noted by Payne (1976). To see this one calculates

$$\frac{\partial P}{\partial n} = 2\left\{\frac{\partial^2 u}{\partial n^2}\frac{\partial u}{\partial n} - \frac{\partial u}{\partial n}\Delta u\right\},$$

and with steps similar to those in Section 5.3, one finds

$$\frac{\partial P}{\partial n} = -2H\left(\frac{\partial u}{\partial n}\right)^2 < 0.$$

Clearly, the assumptions that $H > 0$ and P attains a maximum on ∂D would contradict each other.

Theorem 10.1 can now be extended in various ways. A first generalization is to solutions of

$$\Delta^2 u + \rho(x^1, x^2)f(u) = 0 \qquad \text{in} \quad D. \tag{10.1}$$

The following result is due to Schaefer (1977).

Theorem 10.2 Let u be a sufficiently smooth solution of (10.1) where $\rho > 0$, $\Delta\rho \leq 0$ in $D \subset E^2$, and for any s we have $sf(s) \geq 0$. Then the function

$$P = \frac{1}{\rho}(|\nabla u|^2 - u\,\Delta u)$$

assumes its maximum on ∂D unless $P < 0$ in D.

Proof We calculate as usual

$$P_{,k} = -\frac{\rho_{,k}}{\rho^2}(|\nabla u|^2 - u\,\Delta u) + \frac{1}{\rho}(2u_{,ik}u_{,i} - u_{,k}\Delta u - u(\Delta u)_{,k})$$

$$\Delta P = \Delta\left(\frac{1}{\rho}\right)(|\nabla u|^2 - u\,\Delta u) + 2\nabla\left(\frac{1}{\rho}\right)\cdot\nabla(|\nabla u|^2 - u\,\Delta u)$$
$$+ \frac{1}{\rho}\Delta(|\nabla u|^2 - u\,\Delta u).$$

Combining we have

$$\Delta P - 2\rho\,\nabla P\cdot\nabla\left(\frac{1}{\rho}\right) = \left(\Delta\left(\frac{1}{\rho}\right) - 2\rho\left|\nabla\left(\frac{1}{\rho}\right)\right|^2\right)(|\nabla u|^2 - u\,\Delta u)$$
$$+ \frac{1}{\rho}\Delta(|\nabla u|^2 - u\,\Delta u).$$

Now,

$$\Delta(|\nabla u|^2 - u\,\Delta u) = 2u_{,ik}u_{,ik} - (\Delta u)^2 - u\,\Delta^2 u$$
$$= 2u_{,ik}u_{,ik} - (\Delta u)^2 + \rho u f(u) \geq 0$$

by Lemma 5.4 and our assumptions on ρ and $f(u)$. Furthermore we have

$$\Delta\left(\frac{1}{\rho}\right) - 2\rho\left|\nabla\left(\frac{1}{\rho}\right)\right|^2 = -\frac{1}{\rho^2}\Delta\rho \geq 0$$

by assumption. Thus P satisfies

$$\Delta P - 2\rho\nabla\left(\frac{1}{\rho}\right)\cdot\nabla P + \frac{\Delta\rho}{\rho}P \geq 0 \qquad \text{in} \quad D,$$

and the statement of Theorem 10.2 follows immediately.

Note that if $\rho \equiv \text{const}$, then the condition "unless $P < 0$ in D" in Theorem 10.2 can be omitted.

As a simple application of Theorem 10.2, we consider the problems

$$\begin{aligned} \Delta^2 u + \rho f(u) &= 0 \qquad \text{in} \quad D \subset E^2, \\ u = \frac{\partial u}{\partial n} &= 0 \qquad \text{on} \quad \partial D \end{aligned} \tag{10.2a}$$

and

$$\begin{aligned} \Delta^2 u + \rho f(u) &= 0 \qquad \text{in} \quad D \subset E^2, \\ u = \Delta u &= 0 \qquad \text{on} \quad \partial D. \end{aligned} \tag{10.2b}$$

As was noted by Schaefer (1977) we have the following consequence of Theorem 10.2.

Corollary 10.1 If $\rho > 0$, $\Delta\rho \leq 0$, and $sf(s) \geq 0$, then problems (10.2a) and (10.2b) have no nontrivial solution in a convex plane domain.

Proof By Theorem 10.2 any nonconstant, nonnegative P must assume its maximum on ∂D. But on ∂D, we easily get (case (10.2b))

$$\frac{\partial P}{\partial n} = -2k\left(\frac{\partial u}{\partial n}\right)^2 < 0$$

since the curvature k is supposed to be positive. Hence $P \equiv \text{const}$ or $P < 0$ in D. Now if P is a constant, then $P \equiv 0$, i.e., $|\nabla u| = 0$, so that $u = \text{const} = 0$ is the only possibility then. We are left to check the case that $P < 0$ in D. But $P < 0$ means that

$$|\nabla u|^2 - u\,\Delta u < 0 \qquad \text{in} \quad D,$$

so that

$$u\,\Delta u > 0 \qquad \text{in} \quad D$$

would have to hold. Thus we have the three cases

(a) $u > 0$ in D, $\Delta u > 0$ in D,
(b) $u < 0$ in D, $\Delta u < 0$ in D,
(c) $u < 0$, $\Delta u < 0$ in $D' \subset D$ and $u > 0$, $\Delta u > 0$ in $D \backslash D'$.

Since $u = 0$ on ∂D, the maximum principle implies that $u \equiv 0$ in cases (a) and (b). In case (c) we again use the fact that $u = 0$ on $\partial D'$ also, and $u \equiv 0$ in D' (and D) follows as well.

If u is a solution of (10.2a), we apply Theorem 10.2 directly, which gives

$$|\nabla u|^2 - u\,\Delta u < 0 \qquad \text{in} \quad D,$$

and after integrating over D

$$2\int_D |\nabla u|^2\,dx < 0,$$

which is a contradiction. Hence again we must have

$$u \equiv \text{const} \equiv 0 \qquad \text{in} \quad D.$$

The functions P treated in Theorems 10.1 and 10.2 are closely related to those considered for second-order problems. However, there are various other, yet more complicated, possibilities as illustrated in the next result, which is due to Payne (1976).

Theorem 10.3 Let u be a sufficiently smooth solution of (10.1) with $\rho \equiv 1$ and $f \leq 0$. Set

$$(\Delta u)_M := \max_{\partial D} \Delta u.$$

Then if $D \subset E^N$, the function

$$\begin{aligned} P := u_{,ij}u_{,ij} - \nabla u \cdot \nabla(\Delta u) - \int_0^u f(y)\,dy \\ - \frac{2(N-1)}{N+2}(\Delta u)_M\,\Delta u + \frac{N-4}{2(N+2)}(\Delta u)^2 \end{aligned}$$

attains its maximum value on ∂D.

Proof The Laplacian of P is given by

$$\begin{aligned} \Delta P = {} & 2(\Delta u)_{,ij}u_{,ij} + 2u_{,ijk}u_{,ijk} - |\nabla(\Delta u)|^2 - 2(\Delta u)_{,ij}u_{,ij} \\ & - \nabla u \cdot \nabla(\Delta^2 u) - f(u)\,\Delta u - f'(u)|\nabla u|^2 - \frac{2(N-1)}{N+2}(\Delta u)_M\,\Delta^2 u \\ & + \frac{N-4}{(N+2)}(\Delta^2 u\,\Delta u + |\nabla(\Delta u)|^2) \\ = {} & 2u_{,ijk}u_{,ijk} - \frac{6}{N+2}|\nabla(\Delta u)|^2 - \frac{2(N-1)}{N+2} f(u)[(\Delta u)_M - \Delta u]. \end{aligned}$$

Since $f(u) \leq 0$, it follows from the differential equation (10.1) that Δu is subharmonic, and therefore

$$\Delta u \leq (\Delta u)_M$$

holds in D.

It remains to be shown that the sum of the terms with the third derivatives of u in the expression for ΔP are nonnegative. This is expressed in the following:

Lemma 10.1 Let v be an arbitrary $C^3(D)$ function, $D \subset E^N$. Then

$$v_{,ijk}v_{,ijk} \geq \frac{3}{N+2}|\nabla(\Delta v)|^2. \tag{10.3}$$

The following proof is somewhat shorter than the original one in Payne (1976).

Proof For an arbitrary number γ we have

$$\sum_{i,j,k} [v_{,ijk} - \gamma((\Delta v)_{,i}\delta_{jk} + (\Delta v)_{,j}\delta_{ik} + (\Delta v)_{,k}\delta_{ij})]^2 \geq 0.$$

This inequality reduces to

$$v_{,ijk}v_{,ijk} - 6\gamma|\nabla(\Delta v)|^2 + 3\gamma^2(N+2)|\nabla(\Delta v)|^2 \geq 0,$$

and the discriminant condition for the quadratic expression in γ is equivalent to (10.3). This completes the proof of Theorem 10.3.

Note that in two dimensions P as given in Theorem 10.3 takes the form

$$P = u_{,ij}u_{,ij} - \nabla u \cdot \nabla(\Delta u) - \int_0^u f(y)\,dy - \frac{1}{4}(\Delta u + 2(\Delta u)_M)\,\Delta u. \tag{10.4}$$

Let u be a solution of (10.2a) with $\rho \equiv 1$. Then Theorem 10.3 states that

$$P \leq \max_{\partial D}\,(u_{,ij}u_{,ij} - \tfrac{1}{4}(\Delta u)^2 - \tfrac{1}{2}\Delta u(\Delta u)_M),$$

which we may rewrite as

$$u_{,ij}u_{,ij} - \nabla u \cdot \nabla(\Delta u) - \int_0^u f(y)\,dy - \frac{1}{4}(\Delta u + (\Delta u)_M)^2 \leq \delta^2, \tag{10.5}$$

where $\delta := \max_{\partial D}|\Delta u|^2$. An integration of (10.5), using the fact that

$$\int_D u_{,ij}u_{,ij}\,dx = \int_D (\Delta u)^2\,dx$$

(see below) if $u = (\partial u/\partial n) = 0$ on ∂D, gives

$$\frac{7}{4}\int_D (\Delta u)^2\,dx - \int_D \left(\int_0^u f(y)\,dy\right)\,dx \leq \frac{5}{4}\delta^2 A, \tag{10.6}$$

where A is the area of D. A particularly simple case occurs when $f(u) = -c$, $c > 0$, in which case problem (10.2a) may be interpreted as the equation of a thin elastic plate under a constant load, clamped on the boundary. Then

$$c \int_D u \, dx = \int_D (\Delta u)^2 \, dx,$$

so that (10.6) gives

$$\int_D (\Delta u)^2 \, dx \leq \frac{5}{11} \, \delta^2 A, \tag{10.7}$$

which is a bound for the potential energy of the plate in terms of δ and A.

Let us mention the derivation of the relation

$$\int_D u_{,ij} u_{,ij} \, dx = \int_D (\Delta u)^2 \, dx \qquad \text{if} \quad u = \frac{\partial u}{\partial n} = 0 \qquad \text{on} \quad \partial D.$$

We have

$$\tfrac{1}{2} \Delta(|\nabla u|^2) \, dx = u_{,ij} u_{,ij} + \nabla u \cdot \nabla(\Delta u),$$

and by Green's theorem

$$\frac{1}{2} \int_D \Delta(|\nabla u|^2) \, dx = \frac{1}{2} \oint_{\partial D} \frac{\partial}{\partial n} (|\nabla u|^2) \, dx = \oint_{\partial D} \frac{\partial^2 u}{\partial n^2} \frac{\partial u}{\partial n} \, ds = 0.$$

Hence it follows that

$$0 = \int_D u_{,ij} u_{,ij} \, dx + \int_D \nabla u \cdot \nabla(\Delta u) \, dx = \int_D u_{,ij} u_{,ij} \, dx - \int_D (\Delta u)^2 \, dx.$$

As a last result in the case of Eq. (10.1) with ($\rho \equiv 1$) we mention

Theorem 10.4 (Payne, 1976) Assume that $f'(u) \leq 0$ in $D \subset E^N$. Then the function

$$P := u_{,ij} u_{,ij} - \nabla u \cdot \nabla(\Delta u) + \frac{N-4}{N+2} \int_0^u f(y) \, dy + \frac{N-4}{2(N+2)} (\Delta u)^2$$

takes its maximum value on ∂D.

Proof A straightforward calculation gives

$$\begin{aligned} \Delta P = {} & 2(\Delta u)_{,ij} + 2u_{,ijk} u_{,ijk} - |\nabla(\Delta u)|^2 - \nabla u \cdot \nabla(\Delta^2 u) \\ & - 2(\Delta u)_{,ij} u_{,ij} + \frac{N-4}{N+2} (f(u) \Delta u + f' |\nabla u|^2) \\ & + \frac{N-4}{N+2} (\Delta u \, \Delta^2 u + |\nabla(\Delta u)|^2). \end{aligned}$$

By Lemma 10.1 and the differential equation (10.1) with $\rho \equiv 1$ it follows that

$$\Delta P \geq -\frac{2}{N+2} f'(u)|\nabla u|^2 \geq 0$$

since $f'(u) \leq 0$ in D was assumed. This implies Theorem 10.4.

If u, $\partial u/\partial n = 0$ on ∂D, we thus get

$$P \leq \frac{3N}{2(N+2)} d^2, \tag{10.8}$$

and integration over D now gives the slightly cruder inequality

$$\int_D (\Delta u)^2 \, dx \leq \frac{1}{2} \delta^2 A.$$

But if $N \leq 4$, we may drop the last two terms in the expression for P in Theorem 10.4 to get

$$u_{,ij}u_{,ij} - \nabla u \cdot \nabla(\Delta u) \leq \delta^2. \tag{10.9}$$

Since

$$\Delta(|\nabla u|^2) = 2u_{,ij}u_{,ij} + 2\nabla u \cdot \nabla(\Delta u),$$

(10.9) implies that

$$\Delta(|\nabla u|^2) \geq -2\delta^2 \qquad \text{in} \quad D. \tag{10.10}$$

Let ψ be the solution of

$$\Delta \psi = -1 \qquad \text{in} \quad D, \qquad \psi = 0 \qquad \text{on} \quad \partial D. \tag{10.11}$$

Now $|\nabla u|^2 = 0$ on ∂D; hence by (10.10) and the maximum principle it follows that

$$|\nabla u|^2 \leq 2\delta^2 \psi. \tag{10.12}$$

In particular it follows that

$$\tau^2 = \max_D \; |\nabla u|^2 \leq 2\delta^2 \psi_M \leq \delta^2 \frac{\rho^2}{2}, \tag{10.13}$$

where ρ is the radius of the inscribed circle ($N = 2$) and where we have used inequality (6.13) in the last step. More results and applications of the kind given in this section can be found in the same paper of Payne (1976).

10.2 ELLIPTIC SYSTEMS

We consider the following system of elliptic equations:

$$\Delta u^\alpha + f^\alpha(u^\beta) = 0 \qquad \text{in} \quad D \subset E^N, \tag{10.14}$$

where Greek indices run from 1 to m, and the summation convention will be used for Greek indices as well. The following result is then an extension of Theorem 5.2 (with $g(u) \equiv 1$) to the case of system (10.14).

Theorem 10.5 Let $\{u^\alpha\}$ be a sufficiently smooth solution of (10.14). Suppose that there exists a function $F(u^\alpha)$ such that $\partial F/\partial u^\alpha = f^\alpha$ and that for any vectors ξ^α, the quadratic form

$$Q := \frac{\partial^2 F}{\partial u^\alpha \partial u^\beta} \xi^\alpha \xi^\beta \equiv \frac{\partial f^\alpha}{\partial u^\beta} \xi^\alpha \xi^\beta$$

is negative semidefinite. Then the function

$$P := \nabla u^\alpha \cdot \nabla u^\alpha + \frac{2}{N} F(u^\beta) \tag{10.15}$$

attains its maximum on ∂D.

Proof We simply calculate

$$P_{,k} = 2u^\alpha_{,ik} u^\alpha_{,i} + \frac{2}{N} \frac{\partial F}{\partial u^\beta} u^\beta_{,k}$$

$$\Delta P = 2u^\alpha_{,ik} u^\alpha_{,ik} + 2u^\alpha_{,k} (\Delta u^\alpha)_{,k} + \frac{2}{N} \frac{\partial^2 F}{\partial u^\beta \partial u^\gamma} u^\beta_{,k} u^\gamma_{,k} + \frac{2}{N} \frac{\partial F}{\partial u^\beta} \Delta u^\beta.$$

Applying Lemma 5.4 in the first sum on the right above and the differential equations (10.14), we find

$$\Delta P \geq \frac{2}{N} f^\alpha f^\alpha - 2 \frac{\partial f^\alpha}{\partial u^\beta} \nabla u^\alpha \cdot \nabla u^\beta + \frac{2}{N} \frac{\partial^2 F}{\partial u^\beta \partial u^\gamma} \nabla u^\beta \cdot \nabla u^\gamma - \frac{2}{N} \frac{\partial F}{\partial u^\beta} f^\beta.$$

Because of the definition of F, we therefore arrive at

$$\Delta P \geq 2\left(\frac{1}{N} - 1\right) \frac{\partial f^\alpha}{\partial u^\beta} \nabla u^\alpha \cdot \nabla u^\beta = 2\left(\frac{1}{N} - 1\right) Q(\nabla u^\alpha, \nabla u^\beta) \geq 0$$

since $N \geq 2$ and $Q(\nabla u^\alpha, \nabla u^\beta) \leq 0$ by assumption. This proves Theorem 10.5.

Application Suppose that for $\alpha = 1, \ldots, m$ we have $u^\alpha = 0$ on ∂D, and that the average curvature H of ∂D satisfies $H \geq H_0 > 0$. Then the strong maximum principle implies that

$$\frac{\partial P}{\partial n} > 0 \qquad \text{at} \quad x \quad \text{on} \quad \partial D$$

for nonconstant P, which assumes its maximum at x. But with the same calculations as those in Chapter 5, we find

$$\begin{aligned}\frac{\partial P}{\partial n} &= 2\frac{\partial^2 u^\alpha}{\partial n^2}\frac{\partial u^\alpha}{\partial n} + \frac{2}{N} f^\alpha(0, \ldots, 0)\frac{\partial u^\alpha}{\partial n}\\ &= 2\frac{\partial u^\alpha}{\partial n}\left(-f^\alpha(0, \ldots, 0) - (N-1)H\frac{\partial u^\alpha}{\partial n}\right) + \frac{2}{N} f^\alpha(0, \ldots, 0)\frac{\partial u^\alpha}{\partial n},\end{aligned}$$

and thus at x (assuming that $f^\alpha(0, \ldots, 0) > 0$ for all α)

$$f^\alpha(0, \ldots, 0)|\nabla u^\alpha| > NH|\nabla u^\alpha|\,|\nabla u^\alpha| \geq NH_0|\nabla u^\alpha|\,|\nabla u^\alpha|. \tag{10.16}$$

Inequality (10.16) may be rewritten as

$$\sum_{\alpha=1}^{m}\left(\tau^\alpha - \frac{f^\alpha}{2NH_0}\right)^2 \leq \sum_{\alpha=1}^{m}\left(\frac{f^\alpha}{2NH_0}\right)^2, \tag{10.17}$$

where $\tau^\alpha\tau^\alpha = \max_{\partial D} \nabla u^\alpha \cdot \nabla u^\alpha$ and the functions f^α are to be evaluated at $(0, \ldots, 0)$.

Example Let D be a strictly convex plane domain (curvature $k \geq k_0 > 0$) and consider the system

$$\begin{aligned}\Delta u + f(u) - \gamma(u + v) &= 0\\ \Delta v + g(v) - \gamma(u + v) &= 0\end{aligned} \qquad \text{in} \quad D \tag{10.18}$$

with $u = v = 0$ on ∂D. In this case we may take

$$P = |\nabla u|^2 + |\nabla v|^2 + \int_0^u f(y)\,dy + \int_0^u g(y)\,dy - \frac{\gamma}{2}(u + v)^2,$$

and the corresponding quadratic form becomes

$$Q(\xi, \boldsymbol{\eta}) = (f'(u) - \gamma)\boldsymbol{\xi}^2 + (g'(v) - \gamma)\boldsymbol{\eta}^2 - 2\gamma\boldsymbol{\xi\eta}.$$

Then $Q(\boldsymbol{\xi}, \boldsymbol{\eta})$ is negative semidefinite, provided that

$$f'(u) - \gamma \leq 0, \qquad g'(v) - \gamma \leq 0 \tag{10.19}$$

and

$$(f'(u) - \gamma)(g'(v) - \gamma) \geq \gamma^2.$$

A sufficient condition is therefore

$$\gamma > 0, \qquad f'(u) \le 0, \qquad g'(v) \le 0. \tag{10.20}$$

Suppose then that $f(0) > 0$, $g(0) > 0$, and, e.g., (10.20) is satisfied. Let us set

$$\tau_u^2 + \tau_v^2 := \max_{\partial D} (|\nabla u|^2 + |\nabla v|^2). \tag{10.21}$$

Then inequalities (10.17) in the case of the system (10.18) become

$$\left(\tau_u - \frac{f(0)}{4k_0}\right)^2 + \left(\tau_v - \frac{g(0)}{4k_0}\right)^2 \le \frac{1}{16k_0^2}(f(0)^2 + g(0)^2). \tag{10.22}$$

Remarks (a) Clearly, Theorem 10.5 can immediately be extended to the case in which the Laplacian in (10.14) is replaced by a uniformly elliptic operator by the method given in Chapter 8.

(b) Note that $\max_{\partial D} |\nabla u| \ge \tau_u$ and $\max_{\partial D} |\nabla v| \ge \tau_v$ in the case of system (10.18). On the other hand, in many practical cases it may be more important to have a bound for the maximum of the total flux of material on ∂D, as given by $\tau_u^2 + \tau_v^2$, instead of the maximal fluxes for u and v.

(c) As in Remark (b) on Theorem 10.1, one could show that if $H > 0$ (e.g., ∂D is strictly convex) and $u^\alpha = 0$ on ∂D, then the function

$$P := \nabla u^\alpha \cdot \nabla u^\alpha + 2F(u^\beta), \qquad F \text{ as in Theorem 10.5}, \tag{10.23}$$

cannot assume its maximum on ∂D ($P \not\equiv$ const).

10.3 PARABOLIC SYSTEMS

We are going to prove an extension of Theorem 9.1 (with $h \equiv 1$) that is valid for a special class of parabolic systems. Let $\{u^\alpha\}$ be a sufficiently smooth solution of

$$\begin{aligned} &u^\alpha_{,t} = \Delta u^\alpha + f^\alpha(u^\beta) \quad \text{in} \quad \Omega \times (0, T), \qquad \Omega \subset E^N, \\ &u^\alpha = 0 \quad \text{on} \quad \partial\Omega \times (0, T) \quad \text{for} \quad \alpha = 1, \ldots, m. \end{aligned} \tag{10.24}$$

At this point it is a strong restriction that the diffusion coefficients are all equal (taken to be 1). However one could again take a uniformly elliptic operator L in the place of the Laplacian. One then has

Theorem 10.6 Let $\{u^\alpha\}$ be a solution of (10.24) with Dirichlet boundary conditions and given initial value $\{u_0^\alpha\}$. Suppose that the following conditions are satisfied

(i) $\dfrac{\partial f^\alpha}{\partial u^\beta} \ge 0$ for $\alpha \ne \beta$;

(ii) the quadratic forms

$$Q^{\alpha}(\xi) := \frac{\partial^2 f^{\alpha}}{\partial u^{\beta}\,\partial u^{\gamma}}\,\xi^{\beta}\xi^{\gamma}$$

are negative semidefinite;

(iii) $f^{\alpha}(0, \ldots, 0) \geq 0$, and there exists a constant $a \in (0,1)$ such that for all $u_0^{\alpha}(x)$ we have

$$\Delta u_0^{\alpha} + (1-a)f^{\alpha}(u_0^{\beta}) \leq 0 \qquad \text{in} \quad \Omega.$$

Then in $\Omega \times (0, T)$ we have for $\alpha = 1, \ldots, m$

$$u_{,t}^{\alpha} - af^{\alpha}(u^{\beta}) \leq 0. \tag{10.25}$$

Proof We set

$$v^{\alpha}(x,t) = u_{,t}^{\alpha}(x,t) + h^{\alpha}(u^{\beta}(x,t)). \tag{10.26}$$

A straightforward calculation then shows that

$$\Delta v^{\alpha} - \frac{\partial f^{\alpha}}{\partial u^{\beta}}\,v^{\beta} - v_{,t}^{\alpha} = \frac{\partial f^{\alpha}}{\partial u^{\beta}}\,h^{\beta} - \frac{\partial h^{\alpha}}{\partial u^{\beta}}\,f^{\beta} + \frac{\partial^2 h^{\alpha}}{\partial u^{\beta}\,\partial u^{\gamma}}\,\nabla u^{\beta} \cdot \nabla u^{\gamma}. \tag{10.27}$$

Choosing $h^{\alpha}(u^{\beta}) = -af^{\alpha}(u^{\beta})$, we see that the right-hand side in (10.27) is nonnegative under assumption (ii). If condition (i) is satisfied, then the maximum principle for parabolic systems (see Theorem 2.10) can be applied, telling us that v must assume its maximum on $\partial\Omega \times (0, T)$ or for $t = 0$. But by assumption (iii), $v \leq 0$ on $\partial\Omega \times (0, T) \cup \Omega \times \{0\}$, and therefore $v \leq 0$ in $\Omega \times (0, T)$, which is expressed by inequality (10.25).

Remarks (a) If we choose $v^{\alpha} = h^{\alpha}(u^{\beta}) - u_{,t}^{\alpha} = af^{\alpha}(u^{\beta}) - u_{,t}^{\alpha}$, it is easy to see with the same proof that

$$u_{,t}^{\alpha} - af^{\alpha}(u^{\beta}) \geq 0, \tag{10.28}$$

provided that the corresponding conditions in Theorem 10.5 now read

(ii) Q^{α} is positive semidefinite, $\alpha = 1, \ldots, m$.
(iii) $f^{\alpha}(0, \ldots, 0) \leq 0$, and there exists a constant a such that

$$\Delta u_0^{\alpha} + (1-a)f^{\alpha}(u_0^{\beta}) \geq 0 \qquad \text{in} \quad \Omega, \qquad \alpha = 1, \ldots, m.$$

(b) In the special case that $a = 0$ we find

Corollary 10.2 Let $\{u^{\alpha}\}$ be a sufficiently smooth solution of (10.24) and suppose that

(i) $\dfrac{\partial f^{\alpha}}{\partial u^{\beta}} \geq 0$ for $\alpha \neq \beta$ and

(iii) $\Delta u_0^\alpha + f^\alpha(u_0^\beta) \geq 0\,(\leq 0)$ in Ω.

Then

$$u^\alpha_{,t} \geq 0\,(\leq 0) \qquad \text{in} \quad \Omega \times (0, T). \tag{10.29}$$

Note that assumption (ii) can be dropped since $a = 0$.

Example Consider the system

$$\begin{aligned} u_{,t} &= \Delta u + f(v) \\ v_{,t} &= \Delta v + g(u) \end{aligned} \qquad \text{in} \quad \Omega \times (0, T), \tag{10.30}$$

and $u = v = 0$ on $\partial\Omega \times (0, T)$ with initial data $u_0, v_0 \equiv 0$. Suppose that

$$\frac{df}{dv} > 0, \qquad \frac{dg}{du} > 0 \qquad \text{and} \qquad f(0) > 0, \qquad g(0) > 0.$$

Then the assumptions of Theorem 10.6 are satisfied with $a = 1$, and we have therefore

$$u_{,t} \leq f(v), \qquad v_{,t} \leq g(u) \qquad \text{in} \quad \Omega \times (0, T). \tag{10.31}$$

Also, since $f(0) > 0$, $g(0) > 0$, we have by Corollary 10.1

$$u_{,t} \geq 0, \qquad v_{,t} \geq 0 \qquad \text{in} \quad \Omega \times (0, T). \tag{10.32}$$

Let $\mu(t)$, $\nu(t)$ be the solutions of

$$\begin{aligned} \dot{\mu} &= f(\nu) \\ \dot{\nu} &= g(\mu) \end{aligned} \qquad \mu(0) = \nu(0) = 0. \tag{10.33}$$

Then since

$$\frac{df}{d\nu} \geq 0, \qquad \frac{dg}{d\mu} \geq 0,$$

it follows that

$$u(x, t) \leq \mu(t), \qquad v(x, t) \leq \nu(t) \qquad \text{in} \quad \Omega \times (0, T). \tag{10.34}$$

Concluding remark One could also derive a different type of maximum principle in the case of a system like (10.24). For instance, let u, v be the solutions of

$$\begin{aligned} u_{,t} &= \Delta u + \frac{\partial F(u, v)}{\partial u} \\ v_{,t} &= \Delta v + \frac{\partial F(u, v)}{\partial v} \end{aligned} \qquad \text{in} \quad \Omega \times (0, T), \qquad \Omega \subset E^2. \tag{10.35}$$

Suppose the quadratic form

$$Q(\xi,\eta) := \frac{\partial^2 F}{\partial u^2}\,\xi^2 + 2\,\frac{\partial^2 F}{\partial u\,\partial v}\,\xi\eta + \frac{\partial^2 F}{\partial v^2}\,\eta^2$$

is negative semidefinite $F_u \geq 0$, $F_v \geq 0$, and in Ω we have

$$\Delta u_0 + F_u(u_0, v_0) \geq 0,$$
$$\Delta v_0 + F_v(u_0, v_0) \geq 0.$$

Then the function

$$P = |\nabla u|^2 + |\nabla v|^2 + F(u, v) \tag{10.36}$$

assumes its maximum on ∂D.

The proof consists of an easy combination of Lemma 5.4 and Corollary 10.1.

Chapter 11
ADDITIONAL AND RELATED RESULTS

11.1 LEVEL SURFACE COORDINATES

The basic idea presented in this section has quite a long history, dating back in its roots to the works of Steiner (1882). Much later the book by Pólya and Szegö (1951) gave a thorough discussion of the method of symmetrization and its applications in mathematical physics. The same topic, in an updated and extended version, has been presented in the recent book by Bandle (1980).

Here, we shall follow a slightly different method. It was first exploited (in the Euclidean case), in the same way as will be done here, in the articles of Payne *et al.* (1977) and Sperb and Stakgold (1979).

Let D be a simply connected finite domain in N-dimensional Euclidean space or in an N-dimensional Riemannian manifold $\mathfrak{M}$. In order to simplify the notation we shall use the terminology of a two-dimensional plane domain. So, e.g., we shall denote the N-dimensional volume of D as just the "area" A, the $N-1$-dimensional surface area of ∂D as the "length" L. Analogously, we shall write just Δ for the Laplacian or the Laplace–Beltrami operator of $\mathfrak{M}$.

We now consider an arbitrary smooth positive function u defined in D such that

(a) $u = 0$ on ∂D,
(b) $|\nabla u|$ vanishes only at a finite number of isolated points in D.

Let $D(t)$ be the subdomain of D, where $u > t$ with boundary $\Gamma(t)$ and "area" $a(t)$. The length of $\Gamma(t)$ will be denoted by $l(t)$. Finally, we shall write A for the total area $a(0)$ and L for the total length $l(0)$ of ∂D.

Let n be the unit outward normal on the boundary $\Gamma(t)$ of $D(t)$ and s the element of $\Gamma(t)$. Then

$$du = -|\nabla u|\, dn,$$

and the area between $\Gamma(t)$ and $\Gamma(t + dt)$ can be written as

$$da = -\oint_{\Gamma(t)} \frac{ds}{|\nabla u|}\, dt. \tag{11.1}$$

In formula (11.1) our basic assumption (b) is used. Denoting the maximum value of u in D by u_M, we may write the area enclosed by $\Gamma(t)$ as

$$a(t) = \int_t^{u_M} du \oint_{\Gamma(u)} \frac{ds}{|\nabla u|}. \tag{11.2}$$

Hence one has

$$A = \int_0^{u_M} du \int_{\Gamma(u)} \frac{ds}{|\nabla u|} = \int_D dx.$$

In the same way we see that for an arbitrary function $E(u, |\nabla u|)$ we have

$$\int_D E(u(x), |\nabla u(x)|)\, dx = \int_0^{u_M} \left(\oint_{\Gamma(u)} \frac{E(u, |\nabla u|)}{|\nabla u|}\, ds \right) du.$$

Example The "Dirichlet integral" $\mathscr{D}(u) = \int_D |\nabla u|^2\, dx$ can be written as

$$\mathscr{D}(u) = \int_0^{u_M} \left(\oint_{\Gamma(u)} |\nabla u|\, ds \right) du.$$

We may also consider u as a function of the independent variable a. Then (11.1) shows that

$$\frac{du}{da} = -\left(\oint_{\Gamma(a)} \frac{ds}{|\nabla u|} \right)^{-1}, \tag{11.3}$$

where we denote by $\Gamma(a) = \Gamma(u(a))$ the "level line" enclosing the area a. Finally, for an arbitrary domain Ω we introduce the purely geometric number

$$\gamma_N(\Omega) := \inf_{\tilde{\Omega} \subset \Omega} \frac{l^2(\partial\tilde{\Omega})}{(a(\tilde{\Omega}))^{2-(2/N)}}, \qquad N \geq 2. \tag{11.4}$$

Note that for a plane domain Ω one has $\gamma_2 = 4\pi$, and for any $N \geq 2$ and a domain in Euclidean space it follows from the classical isoperimetric inequality that γ_N is just the value for the N-ball. However, if Ω is a domain in a Riemannian manifold $\mathfrak{M}$, the number γ_N will depend on the metric of $\mathfrak{M}$. In general it will be difficult to determine γ_N, but there are interesting cases in which this is possible.

We have seen that the Dirichlet integral can be written as

$$\mathscr{D}(u) = \int_0^{u_M} \left(\oint_{\Gamma(u)} |\nabla u| \, ds \right) du.$$

We now combine (11.3) with the integrand of $\mathscr{D}(u)$ to the product

$$\begin{aligned} \oint_{\Gamma(u)} |\nabla u| \, ds \left(-\frac{da}{du} \right) &= \oint_{\Gamma(u)} |\nabla u| \, ds \oint_{\Gamma(u)} \frac{ds}{|\nabla u|} \\ &\geq l^2(u) \\ &\geq \gamma_N(D(u))[a(D(u))]^{2-(2/N)}. \end{aligned}$$

If we then choose a as our new independent variable we arrive at the inequality

$$\mathscr{D}(u) \geq \int_0^A \gamma_N(D(a)) a^{2-(2/N)} \left(\frac{du}{da} \right)^2 da \geq \gamma_N(D) \int_0^A a^{2-(2/N)} \left(\frac{du}{da} \right)^2 da. \quad (11.5)$$

For example, if D is a plane domain, (11.5) reduces to

$$\mathscr{D}(u) \geq 4\pi \int_0^A a \left(\frac{du}{da} \right)^2 da. \quad (11.5')$$

Further development then depends on the differential equation satisfied by the given function u. We shall therefore consider different cases separately. An important fact is that a number of interesting bounds for solutions of various problems can be derived, either if γ_N can be determined or if a bound for $|\nabla u|$ in terms of u is known. We shall first consider cases for which no gradient bounds are required.

11.2 ESTIMATES NOT REQUIRING GRADIENT BOUNDS

11.2.1 Surface of Constant Mean Curvature

Let D be a plane domain and let $u(x)$ define a nonparametric surface of constant mean curvature $H > 0$ such that $u = 0$ on ∂D. Thus u is the solution of

$$\operatorname{div}\left(\frac{\nabla u}{\sqrt{1 + |\nabla u|^2}} \right) = -2H \quad \text{in } D, \qquad u = 0 \quad \text{on } \partial D. \quad (11.6)$$

Serrin's (1969) existence criterion requires that the curvature $k(s)$ of ∂D satisfy $k(s) \geq 2H$ for $0 \leq s \leq L$. We shall assume that this criterion is satisfied and set

$$S(t) = \int_t^{u_M} dt \oint_{\Gamma(t)} \frac{\sqrt{1 + |\nabla u|^2}}{|\nabla u|} ds = \int_{D(t)} \sqrt{1 + |\nabla u|^2}\, dx. \tag{11.7}$$

Thus $S(t)$ is the surface area above the interior domain $D(t)$. Now from the differential equation (11.6) and Green's identity it follows that the area enclosed by $\Gamma(t)$ is given by

$$a(t) = \frac{1}{2H} \oint_{\Gamma(t)} \frac{|\nabla u|}{\sqrt{1 + |\nabla u|^2}} ds.$$

An easy combination therefore gives

$$-\frac{dS}{dt} 2Ha(t) = \oint_{\Gamma(t)} \frac{\sqrt{1 + |\nabla u|^2}}{|\nabla u|} ds \oint_{\Gamma(t)} \frac{|\nabla u|}{\sqrt{1 + |\nabla u|^2}} ds \geq \left(\oint_{\Gamma(t)} ds \right)^2 = l^2(t),$$

where Schwarz's inequality was used in the second step. At this point we use the classical isoperimetric inequality in the plane which gives $\gamma_2(D(t)) = 4\pi$ or, equivalently,

$$l^2(t) \geq 4\pi a(t).$$

Hence we find the simple inequality

$$-\frac{dS}{dt} \geq \frac{2\pi}{H},$$

and after an integration

$$S \equiv S(0) \geq \frac{2\pi}{H} u_M. \tag{11.8}$$

Inequality (11.8) is isoperimetric since the equality sign holds if D is a disk. Other inequalities in problem (11.6) which require gradient bounds will be given later in Section 11.3.1.

11.2.2 The "Torsion Problem" on a Manifold and in the Plane

Let u be the solution of

$$\Delta u + 1 = 0 \quad \text{in} \quad D, \qquad u = 0 \quad \text{on} \quad \partial D, \tag{11.9}$$

where D is a plane domain or a domain on a two-dimensional Riemannian manifold. If D is a plane domain, problem (11.9) is the torsion problem for an elastic beam of cross section D, and we therefore call it the "torsion problem on a manifold" if D lies on a manifold. Again, the reader is reminded of the close link between (11.9) and the Poisson problem as illustrated in Section 8.3.1.

Note that in the case of problem (11.9) we have

$$a(t) = \oint_{\Gamma(t)} |\nabla u| \, ds,$$

so that a combination of (11.3) and our last equality gives

$$-a(t)\frac{da}{dt} = \oint_{\Gamma(t)} |\nabla u| \, ds \oint_{\Gamma(t)} \frac{ds}{|\nabla u|} \geq l^2(t), \tag{11.10}$$

where again Schwarz's inequality was used in the second step. Assume now that the Gaussian curvature K_G of $\mathfrak{M}$ in D is bounded above by K_m and

$$K_m A < 4\pi.$$

Then it was shown by Alexandrow (1955) that the following extension of the classical isoperimetric inequality holds: For any subdomain $\Omega \subset D$ one has

$$l^2(\partial\Omega) \geq a(\Omega)(4\pi - K_m a(\Omega)). \tag{11.11}$$

Inequalities (11.10) and (11.11) can be combined now to give

$$\frac{du}{da} + \frac{1}{4\pi - K_m a} \geq 0, \qquad a \in (0, A). \tag{11.12}$$

We remark in passing that inequality (11.12) and its analogs later on will always be integrated. It is therefore sufficient that they hold for almost all values of $a \in (0, A)$, which would allow us to weaken our hypothesis (b) of Section 11.1.

Note that for a plane domain D one has to take $K_m = 0$ in inequality (11.12).

We now integrate (11.12) in different ways. If we integrate (11.12) from $a = 0$ to $a = A$ we are led to

$$u_M \leq \frac{1}{K_m} \log \frac{4\pi}{4\pi - K_m A}, \qquad K_m A < 4\pi, \qquad K_m \neq 0, \tag{11.13}$$

or

$$u_M \leq \frac{A}{4\pi}, \qquad K_m = 0. \tag{11.14}$$

Inequality (11.13) was found by Bandle (1971) and (11.14) is due to Polya and Szegö (1951). The equality sign holds in (11.13) if D is a geodesic disk on a surface of constant Gaussian curvature and holds in (11.14) if D is a disk in the plane.

If we first multiply both sides in (11.12) by $4\pi - K_m a$ and then integrate, we have

$$-4\pi u_M - K_m \int_0^A a \frac{du}{da}\, da + A \geq 0.$$

But since $u(A) = 0$ we have

$$\int_0^A a \frac{du}{da}\, da = \int_0^A u\, da = \int_D u\, dx = \mathscr{D}(u),$$

and thus

$$u_M \leq \frac{1}{4\pi}(A + K_m \mathscr{D}(u)). \tag{11.15}$$

Next we multiply both sides in (11.12) by $u(4\pi - K_m a)$ and integrate afterward. A simple calculation now gives

$$2\pi u_M^2 \leq \mathscr{D}(u) + \frac{1}{2} K_m \int_D u^2\, dx. \tag{11.16}$$

For $K_m = 0$ (11.16) reduces to an inequality of Payne (1967). Finally we multiply both sides in (11.12) by a and integrate to get

$$\mathscr{D}(u) \leq \frac{4\pi}{K_m} \log \frac{4\pi}{4\pi - K_m A} - \frac{A}{K_m}, \tag{11.17}$$

as noted also by Bandle (1971). For $K_m = 0$ (11.17) reduces to the inequality of Saint Venant (1856):

$$\mathscr{D}(u) \leq A^2/8\pi. \tag{11.18}$$

Of course there are still more possible ways to integrate inequality (11.12). In particular some interesting isoperimetric inequalities can be deduced if instead of Alexandrow's inequality (11.11) one uses a bound for $|\nabla u|$.

All inequalities given above are again isoperimetric in the sense that equality holds if D is a geodesic disk on a surface of constant Gaussian curvature (or in the plane).

Inequality (11.12) takes a very simple form due to our differential equation (11.9). But even if the constant 1 in (11.9) is replaced by a term $f(u)$, numerous similar integrations can be performed. We first consider the case in which D is a plane domain, but in which a general nonlinearity $f(u)$ takes the place of the constant 1 in (11.9).

11.2.3 Nonlinear Dirichlet Problem in the Plane

Let u be the solution of

$$\Delta u + f(u) = 0 \quad \text{in} \quad D, \qquad u = 0 \quad \text{on} \quad \partial D, \tag{11.19}$$

where D is a simply connected plane domain and $f(u) \geq 0$ for $u \geq 0$. A simple application of the maximum principle shows that for smooth solutions of (11.19) one has $|\nabla u| > 0$ along a simple closed level line of u. It was shown by Sperb (1975) that under some additional assumptions u may indeed have only one interior point where $\nabla u = 0$ (at the maximum), and thus our assumption (b) is satisfied.

We now introduce the function

$$E(a) := \int_0^a f(u(y))\,dy = \int_{u(a)}^{u_M} f(u)\,du \oint_{\Gamma(u)} \frac{ds}{\nabla u}. \tag{11.20}$$

Then we have

$$\frac{dE}{da} = f(u(a))$$

and

$$\frac{d^2E}{da^2} = \frac{df}{du}\frac{du}{da}.$$

By (11.4) we have

$$\frac{d^2E}{da^2} = -\frac{df}{du}\left(\oint_{\Gamma(a)} \frac{ds}{|\nabla u|}\right)^{-1}.$$

On the other hand, by Green's identity and differential equation (11.19) the relation

$$E(a) = \oint_{\Gamma(a)} |\nabla u|\,ds$$

holds. Another simple combination then gives

$$\frac{du}{da} + \frac{E(a)}{4\pi a} \geq 0, \qquad a \in (0, A), \tag{11.21}$$

which may be written in terms of $E(a)$ as

$$\frac{d^2E}{da^2} + \frac{df}{du}\frac{E(a)}{4\pi a} \geq 0, \qquad a \in (0, A). \tag{11.22}$$

The inequalities derived in the following are all isoperimetric with equality holding if D is a disk.
Let us now multiply (11.21) by

$$4\pi a f(u(a)) = 4\pi a \frac{dE}{da},$$

and set

$$F(u) = \int_0^u f(y)\,dy.$$

Then it readily follows that

$$4\pi \int_0^A a\frac{dF}{da}\,da + \int_0^A E\,\frac{dE}{da}\,da = -4\pi \int_0^A F(a)\,da + \frac{1}{2}\,E^2(A) \leq 0$$

since $E(0) = 0$. The last inequality can now be written as

$$\left(\int_D f(u)\,dx\right)^2 \geq 8\pi \int_D F(u)\,dx, \tag{11.23}$$

as found in Payne *et al.* (1977). If u is the first eigenfunction in the fixed membrane problem (11.23) becomes

$$\left(\int_D u\,dx\right)^2 \geq \frac{4\pi}{\lambda_1} \int_D u^2\,dx. \tag{11.24}$$

This converse of Schwarz's inequality was given by Payne and Rayner (1972) who used a slightly different reasoning. For $f(u) \equiv 1$, (11.23) again reduces to the inequality of Saint Venant (11.18).
Let

$$f(u) = \nu u^{2p+1}, \qquad p = 0, 1, 2, \ldots, \qquad \text{and} \qquad \nu > 0.$$

It follows from a result by Levinson (1962) that problem (11.19) has a positive solution for any $\nu > 0$ with this choice of $f(u)$. In some problems of physical interest (see Crooke and Sperb 1978)) it is important to know the value of ν for which

$$\mathscr{D}(u) = 1.$$

It is not hard to see that there is exactly one such value ν, which we shall call ν_1 in the following. It can be characterized as

$$\nu_1 = \underset{\substack{v=0 \\ \text{on } \partial D}}{\text{Min}} \left((\mathscr{D}(v))^{p+1} \Big/ \int_D v^{2p+2} dx \right).$$

We now go back to (11.21), which we multiply by $4\pi a(du/da)$, observing that this expression is negative. An integration and application of the differential equation then yields

$$\int_0^A 4\pi a\left(\frac{du}{da}\right)^2 da \le \nu_1 \int_0^A u^{2p+2}\, da.$$

Now u is normalized such that $\mathscr{D}(u) = 1$. By (11.5)′ we have

$$\int_0^A 4\pi a\left(\frac{du}{da}\right)^2 da \le \mathscr{D}(u) = 1,$$

and hence

$$\nu_1 = \frac{1}{\int_0^A u^{2p+2}\, da} \ge \frac{\left(\int_0^A 4\pi a(du/da)^2\, da\right)^{p+1}}{\int_0^A u^{2p+2}}$$

$$\ge \operatorname*{Min}_{\substack{v(A)=0\\ v(0)<\infty}} \frac{\left(\int_0^A 4\pi a(dv/da)^2\, da\right)^{p+1}}{\int_0^A v^{2p+2}\, da} =: \hat{\nu}_1. \tag{11.25}$$

It was shown by Crooke and Sperb (1978) that the solution for a disk is radially symmetric. If we now introduce a new variable r by setting

$$a = \pi r^2, \qquad A = \pi R^2,$$

we see that the expression for $\hat{\nu}_1$ becomes

$$\hat{\nu}_1 = \operatorname*{Min}_{\substack{v(A)=0\\ v(0)<\infty}} \frac{\left(\int_0^R (dv/dr)^2 r\, dr\right)^{p+1}}{\int_0^R v^{2p+2} r\, dr},$$

which is just the value for a disk of radius R. Hence inequality (11.25) tells us that for a given area A of D, ν_1 is a minimum for a disk. For $p = 0$ inequality (11.25) reduces to the well-known result of Faber (1923) and Krahn (1924) for the first eigenvalue in the fixed membrane problem.

Remark (a) The essential point in the proof of inequality (11.25) is the existence of a radially symmetric solution if D is a disk. Under this assumption a more general inequality than that of (11.25) was given by Bandle (1973).

(b) If in problem (11.19) we replace $f(u)$ by $\lambda f(u)$, and if $f(0) > 0$, and f is increasing and convex, there is a positive solution only for $\lambda \in (0, \lambda^*)$,

$\lambda^* < \infty$ (see Chapter 3). It was then shown by Bandle and Hersch (1975) that for a given area λ^* is a minimum for a disk.

(c) A possible application of inequality (11.25) is the following. For any function $v \in C^1(D) \cap L^{2p+2}(D)$ that vanishes on ∂D we have

$$\int_D v^{2p+2}\,dx \le \frac{1}{\hat{v}_1}(\mathscr{D}(v))^{p+1},$$

$\hat{v}_1$ being the value for a disk. We list a few approximate values for $\hat{v}_1$:

p	$R^2 v_1$	p	$R^2 v_1$
0	5.78	2	68
1	14	3	235

(d) There are of course a number of nonisoperimetric results that can be derived from the basic differential inequality (11.21). Some results of this kind are mentioned, e.g., in the papers of Payne *et al.* (1977) and Crooke and Sperb (1978).

11.2.4 Eigenvalue Problem on a Manifold

Let D be a finite, simply connected domain on an N-dimensional Riemannian manifold $\mathfrak{M}$ and u a solution of

$$\Delta u + \lambda u = 0 \quad \text{in} \quad D, \qquad u = 0 \quad \text{on} \quad \partial D. \tag{11.26}$$

Here Δ stands for the Laplace–Beltrami operator Δ or for the ordinary Laplacian in the case that $\mathfrak{M}$ is flat. Again, we are interested in lower bounds for the first eigenvalue λ_1. We shall first prove the inequality of Cheeger (1969):

$$\lambda_1 \ge \tfrac{1}{4}h^2(D), \tag{11.27}$$

where the purely geometric quantity $h^2(D)$ is defined in contrast to (11.4) as

$$h(D) = \inf_{\tilde{D} \subset D} \frac{l(\tilde{D})}{a(\tilde{D})}, \tag{11.28}$$

$$h^2(D) = \gamma_\infty(D).$$

We can prove (11.27) as follows. As in (11.20) we set

$$\lambda_1 E(a) = \lambda_1 \int_0^a u(y)\,dy = \oint_{\Gamma(a)} |\nabla u|\,ds,$$

u being the first (positive) eigenfunction of (11.26). Then by (11.1) we get

$$-\lambda_1 \frac{da}{du} E(u) \geq l^2(u),$$

and taking a as our independent variable we see again from (11.28) and our last inequality that

$$\frac{du}{da} + \frac{\lambda_1 E}{h^2(D)a^2} \geq 0, \qquad a \in (0, A), \tag{11.29}$$

or, equivalently,

$$\frac{d^2E}{da^2} + \frac{\lambda_1}{h^2(D)} \frac{E}{a^2} \geq 0, \qquad a \in (0, A). \tag{11.30}$$

Note that $E(0) = 0$ and $dE/da|_{a=0} = u_M > 0$.

We now multiply (11.30) by $E(a)$ and integrate to find

$$\frac{\lambda_1}{h^2(D)} \int_0^A \left(\frac{E}{a}\right)^2 da \geq \int_0^A \left(\frac{dE}{da}\right)^2 da,$$

But Hardy's (1920) inequality gives

$$\int_0^A \left(\frac{dE}{da}\right)^2 da > \frac{1}{4} \int_0^A \left(\frac{E}{a}\right)^2 da,$$

from which (11.27) follows.

Remark In some cases one can give a lower bound for $h(D)$. For example, if the Gaussian curvature K_G in D ($N = 2$) is bounded above by $-\alpha^2$, then it was proven by Yau (1975) that

$$l(\tilde{D}) \geq \alpha a(\tilde{D})$$

for any $\tilde{D} \subset D$. Hence

$$h(D) \geq \alpha,$$

and (11.27) implies then that in this case

$$\lambda_1 \geq \tfrac{1}{4}\alpha^2.$$

We can derive a different inequality for λ_1 if we use the number $\gamma_N(D)$ in the place of $h(D)$. The result is then the following inequality

$$\lambda_1 \geq \frac{A^{-2/N}}{N^2} \gamma_N(D)(j^{(N-2)/2})^2. \tag{11.31}$$

where $j^{(N-2)/2}$ is the first zero of the Bessel function $J_{(N-2)/2}$. The equality sign holds in (11.31) if D is the N-ball in Euclidean space.

In order to prove (11.31) we first note that with our definition of $\gamma_N(D)$ the inequality corresponding to (11.29) now becomes

$$\frac{du}{da} + \frac{\lambda_1}{\gamma_N(D)} a^{(2/N)-2} E \geq 0, \qquad a \in (0, A). \tag{11.32}$$

Multiplying by $-a^{2-(2/N)}\, du/da$ and integrating, we get

$$\lambda_1 \geq \gamma_N(D) \frac{\int_0^A a^{2-(2/N)}(du/da)^2\, da}{\int_0^A u^2\, da}$$

$$\geq \gamma_N(D) \operatorname*{Min}_{\substack{v(A)=0 \\ v(0)<\infty}} \frac{\int_0^A a^{2-(2/N)}(dv/da)^2\, da}{\int_0^A v^2\, da}. \tag{11.33}$$

If we again make a transformation of variables by setting

$$a = \frac{\omega_N}{N} r^N, \qquad A = \frac{\omega_N}{N} R^N,$$

where ω_N is the surface area of the unit N-sphere, we see that the last expression on the right in (11.33) is just the value for λ_1 if D is the N-ball in Euclidean space. This value is known to be

$$(j^{(N-2)/2})^2/R^2,$$

which leads to inequality (11.31).

Remark (a) For $N \to \infty$ in (11.31) we have $\gamma_N(D) \to h^2(D)$, and by well-known properties for Bessel functions of large order (see, e.g., Watson (1944)) one has

$$\lim_{N\to\infty} \frac{1}{N^2} (j^{(N-2)/2})^2 = \frac{1}{4}.$$

Hence Cheeger's inequality (11.27) follows as a limiting case from (11.31).

(b) If D is a domain in Euclidean space, inequality (11.31) reduces to the N-dimensional version of the Faber–Krahn inequality since $\gamma_N(D)$ is then just the value of an N-ball.

(c) If $\mathfrak{M}$ is a compact manifold without boundary, it was shown by Cheng (1976) that the nodal set of the first eigenfunction separates $\mathfrak{M}$ into two simply connected subdomains, i.e., $\mathfrak{M} = D_1 \oplus D_2$. Let $\lambda_1(\mathfrak{M})$ be the first eigenvalue of the Laplace–Beltrami operator of $\mathfrak{M}$. Thus

$$\lambda_1(\mathfrak{M}) = \lambda_1(D_1) = \lambda_1(D_2),$$

where $\lambda_1(D_i)$ is the first eigenvalue in problem (11.28) for $D = D_i$, $i = 1, 2$. By (11.31) we must have

$$\lambda_1(\mathfrak{M}) \geq \frac{1}{N^2}(j^{(N-2)/2})^2 \inf_{D_1 \oplus D_2 = \mathfrak{M}} [\max\{\gamma_N(D_1)A_1{}^{-2/N}, \gamma_N(D_2)A_2{}^{-2/N}\}] \tag{11.34}$$

where A_i is the area of D_i.

For example, if $\mathfrak{M}$ is the two-sphere of radius 1, inequality (11.27) gives

$$\lambda_1(\mathfrak{M}) \geq \tfrac{1}{4},$$

whereas (11.34) gives

$$\lambda_1(\mathfrak{M}) \geq 1.44,$$

while $\lambda_1(\mathfrak{M}) = 2$.

11.3 ESTIMATES BASED ON GRADIENT BOUNDS

In Section 11.2 the key inequality was the classical isoperimetric inequality or an extension of it, such as Alexandrow's inequality. In this section we shall see that the combination of gradient bounds and level surface coordinates allows us to improve some of our previous estimates and also leads to a number of additional results.

11.3.1 Surface of Constant Mean Curvature

Let u again be the solution of (11.6) and $a(t)$ the same quantity as in Section 11.2.1. We now combine $a(t)$ and equation (11.3), which yields

$$-2Ha\frac{da}{du} = \oint_{\Gamma(u)} \frac{|\nabla u|}{\sqrt{1 + |\nabla u|^2}}\, ds \oint_{\Gamma(u)} \frac{ds}{|\nabla u|}.$$

A first possibility to estimate the quantity on the right is to apply Theorem 7.3 as well as the classical isoperimetric inequality in the plane as follows.

Let $q = |\nabla u|^2$ and $q_M = \max_{\partial D} |\nabla u|^2$. Then Theorem 7.3 gives the inequality

$$\frac{1}{\sqrt{1+q}} \geq Hu + \frac{1}{\sqrt{1+q_M}} \qquad \text{in} \quad D,$$

and therefore

$$-2Ha\frac{da}{du} \geq \left(Hu + \frac{1}{\sqrt{1+q_M}}\right) \oint_{\Gamma(u)} |\nabla u|\, ds \oint_{\Gamma(u)} \frac{ds}{|\nabla u|}.$$

Schwarz's inequality and the classical isoperimetric inequality can be applied in the last two terms on the right. This shows that

$$-2Ha\frac{da}{du} \geq \left(Hu + \frac{1}{\sqrt{1+q_M}}\right)4\pi a,$$

i.e.,

$$-\frac{da}{du} \geq 2\pi\left(u + \frac{1}{H\sqrt{1+q_M}}\right). \tag{11.35}$$

We now multiply (11.35) by du/da and integrate from $a = 0$ to $a = A$ to get

$$u_M^2 + \frac{2}{H\sqrt{1+q_M}}u_M \leq \frac{A}{\pi}. \tag{11.36}$$

Note that (11.36) becomes an equality if D is a circle since then $q_M = \infty$. We can continue (11.36) on the left by using inequality (7.46) to bound the term involving q_M. This way one finds

$$u_M^2 + u_M\frac{2(k_0^2 - H^2)^{1/2}}{Hk_0} \leq \frac{A}{\pi},$$

where k_0 is the positive lower bound for the curvature of ∂D. If we solve the last inequality for u_M, we are finally led to

$$u_M \leq \left(\frac{A}{\pi} + \frac{1}{H^2} - \frac{1}{k_0^2}\right)^{1/2} - \left(\frac{1}{H^2} - \frac{1}{k_0^2}\right)^{1/2}. \tag{11.37}$$

A second possibility is to proceed as follows. We write

$$-2Ha\frac{du}{da} = \oint_{\Gamma(u)}\frac{|\nabla u|}{\sqrt{1+|\nabla u|^2}}\,ds\left(\oint_{\Gamma(u)}\frac{ds}{|\nabla u|}\right)^{-1}.$$

Since the right side is an increasing function of $|\nabla u|$, Theorem 7.1 can be used to give an upper bound for $|\nabla u|$:

$$|\nabla u| \leq \{[1 - 2H(u_M - u)]^{-2} - 1\}^{1/2}.$$

Hence we have

$$-2Ha\frac{du}{da} \leq \frac{[1 - 2H(u_M - u)]^{-2} - 1}{[1 - 2H(u_M - u)]^{-1}}, \tag{11.38}$$

where is it of course assumed that inequality (7.49), i.e.,

$$2Hu_M < 1$$

holds. Setting

$$g(a) = 1 - 2H(u_M - u(a))$$

we note that (11.38) can be written as

$$-a\frac{dg}{da} \leq \frac{1-g^2}{g}.$$

Separation of variables and integration then yields

$$\frac{1-g^2(a)}{1-g^2(A)} = \frac{1-[1-2H(u_M-u)]^2}{1-[1-2Hu_M]^2} \leq \left(\frac{a}{A}\right)^2,$$

which can be solved for $u(a)$ giving

$$u(a) \geq u_M - \frac{1}{2H}\left(1-\left(1-\beta\left(\frac{a}{A}\right)^2\right)^{1/2}\right),$$

where

$$\beta = 1-[1-2Hu_M]^2.$$

We now integrate the inequality for $u(a)$ from $a = 0$ to $a = A$. After some routine calculation one arrives at

$$1 + 2H\frac{V}{A} \geq 2Hu_M + \frac{1}{2}\left(\sqrt{1-\beta^2} + \frac{\arcsin\beta}{\beta}\right). \tag{11.39}$$

Since the function of β appearing on the right is decreasing in β we may replace (11.39) by (take $\beta = 1$)

$$1 + 2H\frac{V}{A} \geq 2Hu_M + \frac{\pi}{4}. \tag{11.40}$$

Note that the equality sign holds in (11.40) if D degenerates to a strip. Using arguments in the case of problem (11.6) that are similar to the ones in Section 6.4.3, Payne (1980a) derived among other things the inequality

$$2H\frac{V}{A} \leq \frac{\pi}{4}. \tag{11.41}$$

The optimal domain in (11.41) is again a strip. Note that the combination of (11.40) and (11.41) again gives

$$2Hu_M < 1.$$

Clearly, techniques similar to the ones used above could be employed in other nonlinear problems of divergence form, such as, the problem of a capillary surface treated in Section 7.2.3. Also, various other integrations of the basic differential inequalities could be performed leading to inequalities that relate different quantities of interest in this problem.

11.3.2 Torsion Problem on a Manifold

If u is a solution of (11.9), we write, similar to (11.10),

$$-a\frac{du}{da}=\oint_{\Gamma(u)}|\nabla u|\,ds\left(\oint_{\Gamma(u)}\frac{ds}{|\nabla u|}\right)^{-1}.$$

Since we know an upper bound $B(u)$ for $|\nabla u|^2$, we may replace our last equation by the inequality

$$-a\frac{du}{da}\le B(u). \tag{11.42}$$

According to Corollary 8.2, we may take

$$B(u)=\left(\tau^2+\frac{1}{K}\right)e^{-Ku}-\frac{1}{K},$$

provided that $\mathfrak{M}$ is a two-dimensional manifold whose Gaussian curvature K_G is bounded below by K. If we insert this bound into (11.42), we find after a first integration that

$$u(a)\le\frac{1}{K}\log\left[1+K\tau^2\left(1-\frac{a}{A}\right)\right].$$

A second integration from $a=0$ to $a=A$ then leads to

$$\mathscr{D}(u)\le\frac{A}{K}\left\{\left(1+\frac{1}{K\tau^2}\right)\log(1+k\tau^2)-1\right\}. \tag{11.43}$$

The equality sign holds in (11.43) if D is a geodesic circle on a sphere. Note that for a plane domain ($K=0$) (11.43) reduces to

$$\mathscr{D}(u)\le\tfrac{1}{2}\tau^2A,$$

which is equivalent to inequality (6.6).

Let us mention another inequality, which for the sake of simplicity we only derive in the case $K=0$. For $K=0$ we may rewrite (11.42) as

$$-a\frac{du}{da}\le\tau^2-u.$$

We now multiply by $-du/da$ and integrate from 0 to A to get

$$\int_0^A a\left(\frac{du}{da}\right)^2da\le\tau^2u_M-\frac{1}{2}u_M^2.$$

We make use of the fact that

$$\operatorname*{Min}_{\substack{v(A)=0 \\ v(0)<\infty}} \frac{\int_0^A (v')^2\, da}{\left(\int_0^A v\, da\right)^2} = \frac{2}{A^2}.$$

Thus our last inequality can be continued on the left, and therefore

$$A^{-2}\left(\int_0^A v\, da\right)^2 \leq \tau^2 u_M - \frac{1}{2} u_M^2,$$

i.e.,

$$\mathscr{D}(u) \leq A(\tfrac{1}{2}\tau^2 u_M - \tfrac{1}{4}u_M^2)^{1/2}. \tag{11.44}$$

The equality sign holds in (11.44) if D is a disk.

If we use Corollary 8.3 instead of Corollary 8.2, we can take

$$B(u) = \frac{1}{K}(e^{2K(u_M - u)} - 1),$$

provided that $K_G \geq K > 0$ and the geodesic curvature of ∂D is nonnegative. Insertion of our last bound in (11.42) and integration then show that

$$u(a) \leq u_M + \frac{1}{2K}\log\left(1 - \varepsilon\left(\frac{a}{A}\right)^2\right), \qquad \varepsilon := 1 - e^{-2ku_M}, \tag{11.45}$$

which in turn implies that

$$\mathscr{D}(u) \leq u_M A + \frac{A}{2K}\int_{1-\varepsilon}^1 \log(1 - \varepsilon x^2)\, d\dot{x}. \tag{11.46}$$

The integral on the right could be determined explicitly, however, it leads to a somewhat complicated expression in terms of ε. We leave it therefore in the implicit form given above.

For $K \to 0$, (11.46) reduces to

$$\mathscr{D}(u) \leq \tfrac{2}{3}Au_M, \tag{11.47}$$

which would also follow from inequality (6.11) by a direct integration (note that the constant 1 in (11.9) is replaced by 2 in (6.11). However, different types of bounds can be obtained in problem (11.9) by using level-surface coordinates in the special case $K = 0$. Again, the reader is reminded of the fact that the case $K = 0$ also corresponds to the Poisson problem (8.60) if the source term $\rho(x)$ in (8.60) satisfies

$$\Delta(\log \rho) \leq 0 \qquad \text{in} \quad D.$$

For $K = 0$ inequality (11.45) reduces to

$$u(a) \leq u_M\left(1 - \left(\frac{a}{A}\right)^2\right), \tag{11.48}$$

which immediately implies the L^p estimate ($p > 0$)

$$\int_D u^p\,dx \leq u_M^p A \int_0^1 (1 - y^2)^p\,dy = u_M^p A \frac{(\Gamma(p+1))^2\,2^{-2p}}{\Gamma(2p+2)}. \tag{11.49}$$

The equality sign holds in (11.49) in the limit as D degenerates to a strip.

11.3.3 Fixed Membrane Problem on a Manifold or in the Plane

It is quite obvious that level-surface coordinates and maximum principles can also be combined in order to obtain estimates in problem (11.19), even if D is a domain on a manifold. If D is a plane domain, some results of this kind are given in Payne *et al.* (1977). Here we shall restrict our attention to the eigenvalue problem (11.26).

Let E be defined as in Section 11.2.4. Then we have

$$-\frac{du}{da}\lambda_1 E(u) = \oint_{\Gamma(u)} |\nabla u|\,ds\left(\oint_{\Gamma(u)} \frac{ds}{|\nabla u|}\right)^{-1},$$

u being the first positive eigenfunction with associated eigenvalue λ_1. We may again use an upper bound $B(u)$ for $|\nabla u|^2$. If the Ricci curvature of $\mathfrak{M}$ in D is nonnegative as well as the mean curvature of ∂D, then by Theorem 8.3 a possible choice is

$$B(u) = \lambda_1(u_M^2 - u^2).$$

Thus

$$-\frac{du}{da}E \leq u_M^2 - u^2,$$

from which a multiplication by $u(a) = dE/da$ and an easy integration lead to

$$1 - \left(\frac{u}{u_M}\right)^2 \geq \left(\frac{E(u)}{E(A)}\right)^2. \tag{11.50}$$

Observing again that $u = dE/da$, we can solve (11.50) for E and integrate, thus obtaining

$$\int_0^{E(A)} \frac{dE}{\sqrt{1 - (E/E(A))^2}} \leq u_M A,$$

which is equivalent to

$$\int_D u\,dx \leq \frac{2}{\pi} u_M A. \tag{11.51}$$

In the special case in which D is a domain in Euclidean space, this last inequality was first proven by Payne and Stakgold (1972) (see also Section 6.4.3). There is yet another way of integrating (11.50). We rewrite it as

$$\frac{u}{u_M} \leq \sqrt{1 - \left(\frac{E}{E(A)}\right)^2},$$

from which we see that

$$\int_0^A \left(\frac{u}{u_M}\right)^p u\,da \leq \int_0^{E(A)} \left(1 - \left(\frac{E}{E(A)}\right)^2\right)^{p/2} dE, \qquad p > 0,$$

or equivalently

$$\int_D u^{p+1}\,dx \leq u_M^p 2^p \frac{[\Gamma(\frac{1}{2}p + 1)]^2}{\Gamma(p + 2)} \int_D u\,dx. \tag{11.52}$$

For example, we have

$$\int_D u^2\,dx \leq \frac{\pi}{4} u_M \int_D u\,dx. \tag{11.53}$$

Again, the equality signs hold in (11.51)–(11.53) if D degenerates to a strip or its N-dimensional analog.

Concluding Remarks (a) If $K > 0$, some inequalities can be improved, however, at the cost of considerable computational work. For some bounds in the case of a two-dimensional inhomogeneous membrane, the interested reader is referred to the paper of Sperb and Stakgold (1979).

(b) Inequalities similar to (11.51)–(11.53) could be derived in the same way for other problems. For example, if $f(u) = \nu u^q$ in problem (11.19) one could show that

$$\int_D u^q\,dx/u_M^q A \leq 2^{2q/(q+1)}\Gamma\left(\frac{2}{q+1}\right)\bigg/\left(\Gamma\left(\frac{1}{q+1}\right)\right)^2,$$

which contains (11.51) as a special case.

(c) It is also possible to use level lines in a similar way in the case of parabolic problems. Such an extension was given by Bandle (1976).

(d) Another interesting extension was given by Kohler (1975). She used the level lines of the torsion function in connection with the fixed membrane problem to prove a conjecture by Pólya and Szegö.

(e) For a number of other generalizations, including different boundary conditions, the reader is referred to the book by Bandle (1980).

11.4 MAXIMUM PRINCIPLES INVOLVING HARMONIC FUNCTIONS

11.4.1 Maximum Principles

It follows from Theorem 5.1 that if h is a harmonic function in a plane domain, then the function

$$P := (f(h))^{-1}|\nabla h|^2 \tag{11.54}$$

assumes its maximum value on ∂D if $(\log f)'' \leq 0$. The same is true under a different condition on $f(h)$ if $D \subset E^N$. This result can be generalized in the sense that two different harmonic functions, say h and H, are involved. In fact, Payne and Philippin (1979a) established the following result.

Theorem 11.1 Let H, h be two harmonic functions in $D \subset E^N$, with $H \in C^1(\bar{D})$ and $h \in C^0(\bar{D})$, and let $f(h)$ be a positive C^2 function over the range of admissible values of h. Suppose that on this range f satisfies

(i) $(f^{(N-2)/2(N-1)})'' \leq 0$, $N \geq 3$ or
(i′) $(\log f)'' \leq 0$, $N = 2$.

Then the function

$$P = \frac{|\nabla H|^2}{f(h)} \tag{11.55}$$

attains its maximum on ∂D.

It is convenient to prove first

Lemma 11.1 Let $v \in C^2(D)$ and $w \in C^1(D)$. Then at points in D where $|\nabla v|$ and $|\nabla w|$ do not vanish, the following inequality holds ($N \geq 3$)

$$\begin{aligned} v_{,ij}v_{,ij} \geq {} & \frac{1}{N-1}\left(\frac{v_{,ik}v_{,i}v_{,k}}{|\nabla v|^2}\right)^2 + \frac{v_{,ik}v_{,k}v_{,ij}v_{,j}}{|\nabla v|^2} \\ & + \frac{1}{2}\frac{(v_{,ik}v_{,i}w_{,k})^2}{|\nabla v|^2|\nabla w|^2} + \frac{1}{2}\frac{(v_{,ik}v_{,i}w_{,k})^2(v_{,s}w_{,s})^2}{|\nabla v|^4|\nabla w|^4} \\ & - \frac{v_{,ik}v_{,i}w_{,k}v_{,jr}v_{,j}v_{,r}v_{,m}w_{,m}}{|\nabla v|^4|\nabla w|^2} - \frac{2}{N-1}\frac{\Delta v\, v_{,ik}v_{,i}v_{,k}}{|\nabla v|^2}. \end{aligned} \tag{11.56}$$

Proof of Lemma 11.1 We define the quantity χ_{ij} by

$$\begin{aligned} \chi_{ij} = v_{,ij} - \frac{v_{,ik}v_{,k}v_{,j}}{|\nabla v|^2} + \frac{1}{N-1}\frac{v_{,ks}v_{,k}v_{,s}}{|\nabla v|^2}\left\{\delta_{ij} - \frac{v_{,i}v_{,j}}{|\nabla v|^2}\right\} \\ + \frac{1}{2}\frac{v_{,kl}v_{,k}w_{,l}}{|\nabla v|^2|\nabla w|^2}\{w_{,i}v_{,j} - w_{,j}v_{,i}\}, \end{aligned} \tag{11.57}$$

where δ_{ij} is the Kronecker symbol. Inequality (11.56) follows immediately from the fact that $\chi_{ij}\chi_{ij} \geq 0$.

Proof of Theorem 11.1 By differentiation we find

$$P_{,k} = 2f^{-1}H_{,ik}H_{,i} - f^{-2}f'H_{,i}H_{,i}h_{,k}, \tag{11.58}$$

and

$$\Delta P = 2f^{-1}H_{,ik}H_{,ik} - 4f^{-2}H_{,i}H_{,ik}h_{,k} - (f^{-2}f')'H_{,i}H_{,i}h_{,k}h_{,k}. \tag{11.59}$$

At this point we make use of Lemma 11.1 with $v = H$ and $w = h$ to get

$$\begin{aligned}\Delta P \geq f^{-1}\Bigg\{&\frac{2}{N-1}\left(\frac{H_{,ik}H_{,i}H_{,k}}{|\nabla H|^2}\right)^2 + 2\frac{H_{,ik}H_{,k}H_{,ij}H_{,j}}{|\nabla H|^2}\\ &+\frac{(H_{,ik}H_{,i}h_{,k})^2}{|\nabla H|^2|\nabla h|^2} + \frac{(H_{,ik}H_{,k}h_{,i})^2(H_{,s}h_{,s})^2}{|\nabla H|^4|\nabla h|^4}\\ &-2\frac{H_{,ik}H_{,i}h_{,k}H_{,jr}H_{,j}H_{,r}H_{,m}h_{,m}}{|\nabla H|^4|\nabla h|^2}\Bigg\} - 4f'f^{-2}H_{,i}H_{,ik}h_{,k}\\ &-(f^{-2}f')'|\nabla H|^2|\nabla h|^2.\end{aligned} \tag{11.60}$$

The following identities may now be inserted

$$H_{,ik}H_{,i}H_{,k} = \frac{f^{-1}f'}{2}|\nabla H|^2H_{,k}h_{,k} + \frac{f}{2}P_{,k}H_{,k}, \tag{11.61}$$

$$H_{,ik}H_{,i}h_{,k} = \frac{f^{-1}f'}{2}|\nabla H|^2|\nabla h|^2 + \frac{f}{2}P_{,k}h_{,k}, \tag{11.62}$$

$$\begin{aligned}H_{,ik}H_{,k}H_{,ij}H_{,ij} = &\frac{f^2}{4}|\nabla P|^2 + \frac{f'}{2}|\nabla H|^2h_{,k}P_{,k}\\ &+\frac{f^{-2}(f')^2}{4}|\nabla H|^4|\nabla h|^2.\end{aligned} \tag{11.63}$$

This allows to rewrite (11.60) in the form

$$\begin{aligned}\Delta P + w_kP_{,k} \geq &\frac{3-N}{4(N-1)}(f')^2f^{-3}(H_{,k}h_{,k})^2\\ &+\left(\frac{3}{4}f'^2 - ff''\right)f^{-3}|\nabla H|^2|\nabla h|^2,\end{aligned} \tag{11.64}$$

where the term w_k includes the terms involving first derivatives of P. One can easily check that w_k may be unbounded at points where $\nabla H = 0$, but remains bounded at points at which $\nabla h = 0$ if $\nabla H \neq 0$ there. Finally, since

$N \geq 3$ and

$$|\nabla H|^2 |\nabla h|^2 \geq (H_{,i} h_{,i})^2,$$

we see that the right-hand side in (11.64) will be positive, provided that

$$ff'' - \frac{N}{2(N-1)} (f')^2 \leq 0,$$

which is equivalent to assumption (i). Hence we have

$$\Delta P + w_{,k} P_{,k} \geq 0$$

and by Hopf's maximum principle P must assume its maximum on ∂D or at a point where $\nabla H = 0$. But if $|\nabla H| \equiv 0$ then $H = \text{const}$, and Theorem 11.1 is then trivial. In the case $N = 2$ we may use the identity (5.15). An analogous computation now gives

$$\Delta P - \frac{f|\nabla P|^2}{|\nabla H|^2} = -\frac{1}{f} (\log f)'' |\nabla H|^2 |\nabla h|^2.$$

This proves Theorem 11.1.

Remarks (a) It follows from the proof that in the case $N = 2$, P as given by (11.55) satisfies a minimum principle if $(\log f)'' \geq 0$. If $(\log f)'' = 0$, i.e., $f(h) = e^{ah+b}$, then P will assume its maximum and minimum on ∂D.

(b) There are other combinations of harmonic functions and their gradients that satisfy a maximum principle. For example, if the h_i, $i = 0, 1, \ldots, m$, are harmonic and $h_0 > 0$ in D, then it can be shown that the quantity $(h_i h_i)/(h_0^2)$ also assumes its maximum on ∂D.

11.4.2 Applications[†]

11.4.2.1 Bounds for Derivatives of Green's Function

Let $G(x, y)$ be the Green's function for the Laplace equation in D, with singularity at $x = y$ and vanishing for $x \in \partial D$. $G(x, y)$ may be represented as

$$G(x, y) = \frac{|x - y|^{2-N}}{(N-2)\omega_N} - g(x, y), \qquad N > 3, \tag{11.65}$$

where $|x - y| = r$ is the distance between x and y and

$$\omega_N = 2\pi^{N/2} \Gamma\left(\frac{N}{2}\right)$$

[†] See Payne and Philippin (1979a).

is the surface area of the unit sphere E^N. The regular part $g(x, y)$ therefore is (for fixed y) a solution of

$$\begin{aligned} \Delta g &= 0 && \text{in} \quad D, \\ g &= \frac{r^{2-N}}{(N-2)\omega_N} && \text{on} \quad \partial D. \end{aligned} \tag{11.66}$$

If D is a plane domain, we have

$$G(x, y) = \frac{1}{2\pi} \log \frac{1}{r} - g(x, y),$$

so that $g(x, y)$ satisfies

$$\begin{aligned} \Delta g &= 0 && \text{in} \quad D, \\ g &= \frac{1}{2\pi} \log \frac{1}{r} && \text{on} \quad \partial D. \end{aligned} \tag{11.66'}$$

Theorem 11.1 may now be applied with the following choices (y fixed)

$$H(x) = G(x, y), \qquad h = r^{-(N-2)}, \qquad f(h) = \begin{cases} h^{2(N-1)/(N-2)}, & N \geq 3, \\ r^{-2}, & N = 2. \end{cases} \tag{11.67}$$

By Theorem 11.1 $P := r^{2N-2}|\nabla G|^2$ takes its maximum on ∂D or at the singular point $x = y$. However, we shall see that unless D is the N-sphere and y is the center the maximum of P must occur on ∂D. To this end we first rewrite P as

$$P = \omega_N^{-2} + 2\omega_N^{-1} r^{N-2} x^i g_{,i} + r^{2(N-1)}|\nabla g|^2. \tag{11.68}$$

By Theorem 11.1 we have

$$P \leq \max\left\{\max_{\partial D} r^{2(N-1)}\left(\frac{\partial G}{\partial n}\right)^2, \omega_N^{-2}\right\}. \tag{11.69}$$

It suffices to show that

$$\omega_N^{-2} \leq \max_{\partial D}\left(r^{2(N-1)}\left(\frac{\partial G}{\partial n}\right)^2\right), \tag{11.70}$$

where the equality sign holds in (11.70) if and only if $g \equiv$ const, i.e., if D is the N-sphere with center at y. In order to demonstrate the validity of (11.70), we show that $P > \omega_N^{-2}$ in a neighborhood of y. Let K_ε be a ball of radius ε with center at y and $K_\varepsilon \subset D$. Since the last term in (11.68) is nonnegative everywhere, we only need to show that for $g \equiv$ const the second term on the right in (11.68) changes sign in K_ε. This follows from the fact that for any $\rho \in (0, \varepsilon)$ we have

$$\oint_{\partial K_\rho} r^{N-2} x^i g_{,i}\, ds = \rho^{N-1} \oint_{\partial K_\rho} \frac{\partial g}{\partial n}\, ds = \rho^{N-1} \int_{K_\rho} \Delta g\, dx = 0. \tag{11.71}$$

Thus unless $g \equiv \text{const}$, the quantity $r^{N-2}x^i g_{,i}$ must change sign in K_ε. Hence it follows that

$$P \leq \max_{\partial D}\left\{r^{2(N-1)}\left(\frac{\partial G}{\partial n}\right)^2\right\}, \tag{11.72}$$

with equality if $g \equiv \text{const}$ in K_ε (and therefore in D). Let $D \subset \tilde{D}$, and suppose ∂D and $\partial\tilde{D}$ have a common point x. It is a simple consequence of the maximum principle that at x we have

$$\left|\frac{\partial G_D}{\partial n}\right| \leq \left|\frac{\partial G_{\tilde{D}}}{\partial n}\right|. \tag{11.73}$$

If D is convex, then a convenient choice of $\tilde{D}$ is the half-space containing D with common boundary point x. This choice yields the estimate

$$\left|\frac{\partial G_D}{\partial n}\right| \leq \frac{2n_r}{\omega_{N^r}N - 1}, \qquad N \geq 2, \tag{11.74}$$

n_r being the projection of the unit outward normal n at x onto the straight line joining x and y. Since P takes its maximum on ∂D we have from (11.74)

$$|\nabla G| \leq \frac{2}{\omega_N} r^{1-N}, \qquad N \geq 2. \tag{11.75}$$

This inequality holds for convex D at any point $x \neq y$ in D. Finally, as in Chapter 6 we may use (11.75) to obtain a pointwise bound for $G(x, y)$ by integrating (11.75) along a ray.

Let d be the distance from y to the point $\hat{x}$ where the ray from y to x intersects ∂D. Integrating (11.75) along this ray, we find for convex D

$$G(x, y) \leq \begin{cases} \dfrac{2}{(N-2)\omega_N}\left[r^{2-N} - d^{2-N}\right], & N \geq 3, \\[2ex] \dfrac{1}{\pi}\log\dfrac{d}{r}, & N = 2. \end{cases} \tag{11.76}$$

11.4.2.2 Bounds for Electrostatic Capacity and Charge Density

The following is a classical problem of electrostatics in E^3. One seeks the solution u of

$$\begin{aligned} \Delta u &= 0 \quad \text{in} \quad D^* := E^3 - D, \\ u &= 1 \quad \text{on} \quad \partial D, \quad u = 0\left(\frac{1}{r}\right) \quad \text{as} \quad r \to \infty, \end{aligned} \tag{11.77}$$

where r measures the distance from some origin in D. Here u is the electrostatic potential of the conductor, and the charge density on ∂D is given by $|\nabla u|$. According to Theorem 11.1 the function

$$P = \frac{|\nabla u|^2}{u^4}, \tag{11.78}$$

must assume its maximum on ∂D or at inifinity. Using the well-known expansion of u for large r, an easy computation gives

$$\lim_{r \to \infty} P(r) = C^{-2}, \tag{11.79}$$

where C is the capacity of the conductor. Therefore,

$$P \leq \max\left\{\max_{\partial D} P, C^{-2}\right\}. \tag{11.80}$$

We now show that

$$C^{-2} \leq \max_{\partial D} P, \tag{11.81}$$

with equality as before if D is a sphere. To prove (11.81) we consider the level surface where $u = \hat{u}$, $\hat{u} \in (0, 1)$. From the expansion of u in the neighborhood of infinity, we know that for sufficiently small $\hat{u}$ the surface $u = \hat{u}$ will be star shaped with respect to the origin. Let $D(\hat{u})$ be such a star-shaped surface, and $D^*(\hat{u})$ the region exterior to $\partial D(\hat{u})$. The identity (3.17) yields

$$\begin{aligned} 0 = \int_{D^*(\hat{u})} x^i u_{,i} \Delta u \, dx = \oint_{\partial D(\hat{u})} x^i u_{,i} \frac{\partial u}{\partial n} \, ds + \frac{1}{2} \int_{D^*(\hat{u})} |\nabla u|^2 \, dx \\ - \frac{1}{2} \oint_{\partial D(\hat{u})} x^i n_i |\nabla u|^2 \, ds, \end{aligned} \tag{11.82}$$

which leads to

$$4\pi C \hat{u} = \hat{u} \oint_{\partial D(\hat{u})} \frac{\partial u}{\partial n} \, ds = - \oint_{\partial D(\hat{u})} x^i n_i \left(\frac{\partial u}{\partial n}\right)^2 ds. \tag{11.83}$$

Here n points outward from $D^*(\hat{u})$, i.e., $x^i n_i$ is negative. Hence we have

$$\frac{4\pi C}{\hat{u}^3} = - \oint_{\partial D(\hat{u})} x^i n_i P \, ds \leq 3\hat{V} \max_{\partial D(\hat{u})} P, \tag{11.84}$$

where $\hat{V}$ is the volume of $D(\hat{u})$. By the classical Poincaré inequality (see Pólya and Szegö, 1951), one has

$$\hat{V} \leq \tfrac{4}{3}\pi \hat{C}^3, \tag{11.85}$$

where C is the capacity of the condenser with boundary $\partial D(\hat{u})$.

To relate $\hat{C}$ and C we note that since $\partial D(\hat{u})$ is a level surface of u, we have

$$C = \hat{C}\hat{u}, \tag{11.86}$$

and with our previous estimates we now see that

$$C^{-2} \leq \max_{\partial D(u)} P. \tag{11.87}$$

Again, the equality sign can hold for any $\partial D(\hat{u})$ if and only if D is a sphere. Since (11.87) shows that there exist points where $P \geq C^{-2}$ we have established (11.81).

Once more, useful information can be deduced from the application of the strong maximum principle. At a point x_0 on ∂D where P assumes its maximum we know that (unless $P \equiv$ const, i.e., D is a sphere) we have

$$\frac{\partial P}{\partial n} = 2\frac{\partial u}{\partial n}\frac{\partial^2 u}{\partial n^2} - 4\left(\frac{\partial u}{\partial n}\right)^3 > 0, \tag{11.88}$$

and using the differential equation on ∂D, we obtain

$$0 < \frac{\partial u}{\partial n} < H(x_0), \tag{11.89}$$

where H is the average curvature of ∂D. Setting $H_m := \max_{\partial D} H$, we have by (11.81) and (11.89)

$$C \geq H_m^{-1}, \tag{11.90}$$

with equality for a sphere. More isoperimetric inequalities can be deduced by integrating the inequality

$$P(x) \leq H_m^2 \tag{11.91}$$

in various ways. Taking the square root in (11.91) and integrating over ∂D, one is led to

$$4\pi C \leq H_m S, \tag{11.92}$$

where S is the surface area of ∂D. If D is star shaped, i.e., $x^i n_i \leq 0$, a multiplication of (11.91) by $-x^i n_i$ and integration over uD now yield

$$4\pi C \leq 3VH_m^2, \tag{11.93}$$

with equality for a sphere. Finally, an integration of $|\nabla u| \leq H_m u^2$ along a straight line from any point $x \in D^*$ to the nearest point on ∂D leads to

$$u(x) \geq (1 + H_m d)^{-1}, \tag{11.94}$$

where d is the distance from x to ∂D.

Other inequalities can be found in the original article of Payne and Philippin (1979a).

Remark It is quite obvious that the methods of Chapter 8 could be used to extend some of the results of this section. Instead of harmonic function solutions h of

$$Lh = 0$$

may be considered then, L being a uniformly elliptic operator.

REFERENCES

Alexandrow, A. D. (1955). "Die innere Geometrie der konvexen Flächen." Springer-Verlag, Berlin and New York.

Amann, H. (1976b). Fixed point equations and nonlinear eigenvalue problems in ordered Banach spaces, *SIAM Rev.* **18**, 620–709.

Amann, H. (1976c). "Nonlinear Elliptic Equations with Nonlinear Boundary Conditions." New Developments in Differential Equations, North Holland Math. Studies, Vol. 21, pp. 43–63. North-Holland Publ., Amsterdam.

Amann, H. (1976a). Existence and mulitiplicityly theorems for semi-linear elliptic boundary value problems, *Math. Z*, **150**, 281–295.

Amann, H. (1978). Invariant sets and existence theorems for semilinear parabolic and elliptic systems, *J. Math. Anal. Appl.* **65**, 432–467.

Amann, H., and Laetsch, T. H. (1976). Positive solutions of convex nonlinear eigenvalue problems, *Indiana Univ. Math. J.* **25**, 259–270.

Anderson, N., and Arthurs, A. M. (1972). Upper and lower bounds for the torsional stiffness of a prismatic bar, *Proc. Roy. Soc. London Ser. A* **328**, 295–299.

Ball, J. M. (1977). Remarks on blow-up and nonexistence theorems for nonlinear evolution equations, *Quart. J. Math. Oxford Ser.* (2), 473–486.

Bandle, C. (1971). Konstruktion isoperimetrischer Ungleichungen der mathematischen Physik aus solchen der Geometrie, *Comment. Math. Helv.* **46**, 182–213.

Bandle, C. (1973). The Rayleigh-Faber-Krahn theorem for the characteristic values associated with a class of nonlinear boundary value problems, *SIAM. Math. Anal.* **4,** 8–14.

Bandle, C. (1975). Some norm estimates for the solutions of a nonlinear diffusion problem, *Z. Angew Math. Phys.* **26**, 357–359.

Bandle, C. (1976). Isoperimetric inequalities for a class of nonlinear parabolic equations, *Z. Angew. Math. Phys.* **27**, 377–384.

Bandle, C. (1979). On isoperimetric gradient bounds for Poisson problems and problems of torsional creep, *Z. Angew. Math. Phys.* **30**, 713–715.

Bandle, C. (1980). "Isoperimetric Inequalities and Applications." Pitman, London.

Bandle, C., and Hersch, J. (1975). Problemes de Dirichlet non linéaires: Une condition suffisante isopérimétrique pour l'existence d'une solution, *C. R. Acad. Sci. Paris* Ser A-B **280**, 1057–1060.

Berger, M., Gauduchon P. and Mazet, E. (1971). "Le Spectre d'une Variété Riemanienne." Springer Lecture Notes in Math., Vol. 194. Springer-Verlag, Berlin and New York.

Bernstein, S. (1904). Sur la nature analytique des solutions de certaines équationes aux derivées partielles du second ordre, *Math. Ann.* **59**, 20–76.

Bernstein, S. (1926). Sur la nature analytique des solutions des équationes aux dérivées partielles du type elliptique, *Math. Z.* **25,** 505–513.

Bernstein, S. (1927). Ueber ein geometrisches Theorem und seine Anwendungen auf die partiellen Differentialgleichungen vom elliptischen Typus, *Math. Z.* **26**, 551–588.

Chavel, I., and Feldman, E. A. (1977). An optimal Poincaré inequality for convex domains of non-negative curvature, *Arch. Rational Mech. Anal.* **65**, 263–273.

Cheeger, J. (1969). A Lower Bound for the Smallest Eigenvalue of the Laplacian. Problems in Analysis. Symposium in Honor of S. Bochner, Princeton Univ., Princeton, New Jersey.

Cheng S. Y. (1975). Eigenvalue comparison theorems and its geometric applications, *Math. Z.* **143**, 289–297.

Cheng, S. Y. (1976). Eigenfunctions and nodal sets, *Comment. Math. Helv.* **51**, 43–55.

Clement, P. H., and Peletier, L. A. (1970). An anti-maximum principle for second order elliptic operators, *J. Differential Equations* **34**, 219–229.

Cohen, D. S. (1971). Generalized radiation cooling of a convex Solid, *J. Math. Anal. Appl.* **35**, 503–511.

Concus, P., and Finn, R. (1969). On the behavior of a capillary surface in a wedge, *Appl. Math. Sci., Proc. Nat. Acad. Sci. U.S.A.* 292–299.

Concus, P., and Finn, R. (1974a). On capillary free surfaces in the absence of gravity, *Acta Math.* **132**, 177–198.

Concus, P., and Finn, R. (1974b). On capillary free surfaces in a gravitational field *Acta Math.* **132**, 207–223.

Courant, R., and Hilbert, D. (1953). "Methods of Mathematical Physics," Vol. 1. Wiley, New York.

Crooke, P. S., and Sperb, R. P. (1978). Isoperimetric inequalities in a class of nonlinear eigenvalue problems, *SIAM J. Math. Anal.* **9**, 671–681.

Czischke, J. (1978) Semilineare parabolische Differentialgleichungen mit nichtlinearen Randbedingungen. Ph.D. Dissertation, Univ. of Bochum, Germany.

Emmer, M. (1973). Esistenza, unicità e rigolarità delle superfici di quilibrio nei capillari, *Ann. Univ. Ferrara Sez. VII (N.S.)* **18**, 79–94.

Faber, G. (1923). Beweis, dass unter allen homogenen Membranen von gleicher Fläche und Spannung die kreisförmige den tiefsten Grundton gibt, *Sitz. der. bayr. Akad. Wiss.*, 169–172.

Friedman, A.(1963). "Generalised Functions and Partial Differential Equations." Prentice-Hall, Englewood Cliffs, New Jersey.

Friedman, A. (1964). "Partial Differential Equation of Parabolic Type." Prentice-Hall, Englewood Cliffs, New Jersey.

Friedman, A. (1969). "Partial Differential Equations." Holt, New York.

Friedrichs, K. (1928). Rand-und Eigenwertprobleme aus der Theorie der elastischen Platten, *Math. Ann.* **98**, 205–247.

Gilbarg, D., and Trudinger, N. S. (1977). "Elliptic Partial Differential Equations of Second Order." Springer-Verlag, Berlin and New York.

Hadamard, J. (1932). "Le problème de Cauchy et les équations aux dérivées partielles linéaires hiperboliques." Ed. Hermann, Paris.

Hardy, G. H. (1920). Note on a theorem of Hilbert, *Math. Z.* **6**, 345–358.

Hersch, J. (1960) Sur la fréquence fondamentale d'une membrane vibrante: Evaluation par défaut et principe de maximum, *Z. Angew. Math. Mech.* **11**, 387–413.

Hopf, E. (1931). Ueber den funktionalen, insbesondere den analytischen Charakter, der Lösungen elliptischer Differentialgleichungen zweiter Ordnung, *Math. Z.* **34**, 194–233.

Hudjaev, S. J. (1964). Boundary value problems for certain quasilinear elliptic equations, *Soviet Math. Dokl.* **5**, 188–192.

Kazdan, J. L., and Warner, F. W. (1975). Remarks on some quasilinear elliptic equations, *Comm. Pure Appl. Math.* **28**, 567–597.

Kohler, M.-TH. (1975). Démonstration de l'inégalité isopérimétrique $P\lambda^2 \geq \pi j_0^4/2$, conjecturée par Polya et Szegö, *C. R. Acad. Sci. Paris Sér A-B*, **281**, 119–121.

Krahn, E. (1924). Ueber eine von Rayleigh formulierte Minimaleigenschaft des Kreises, *Math. Ann.* **94**, 97–100.

Ladyshenskaya, O. A., and Ural'Ceva (1968). "Linear and Quasilinear Elliptic Equations." Academic Press, New York.

Laetsch, T. H. (1970). The number of solutions of a nonlinear two point boundary value problem, *Indiana Univ. Math. J.* **20**, 1–13.

Langenbach, A. (1964). Verallgemeinerte und exakte Lösungen des Problems der elastisch-plastischen Torsion von Stäben, *Math. Nachr.* **28**, 219–234.

Levinson, N. (1962). Positive eigenfunctions for $\Delta u + \lambda f(u) = 0$, *Arch. Rational Mech. Anal.* **11**, 1065–1072.

McKean, H. P. (1970). An upper bound to the spectrum of Δ on a manifold of negative curvature, *J. Differential Geom.* **4**, 359–366.

Miranda, C. (1948). Formule di maggiorazione e teorema di esistenza per le funzioni biarmoniche di due variabli, *Giorn. Mat. Battaglini* (6) **78**, 97–118.

Miranda, C. (1970). "Partial Differential Equations of Elliptic Type." Springer-Verlag, Berlin and New York.

Murray, J. D., and Sperb, R. P. (1981-a) Diffusion driven instability on growing membrane surfaces. Research report, University of Oxford.

Murray, J. D., and Sperb, R. P. (1981-b) Minimum domains for spatial patterns in a class of reaction diffusion equations, to appear in *SIAM J. Appl. Math.*

Opial, Z. (1961). Sur les périodes des solutions de l'équation $x'' + g(x) = 0$, *Ann. Polon. Math.* **10**, 49–72.

Osserman, R. (1978). The isoperimetric inequality, *Bull. Amer. Math. Soc.* **84**, 1182–1238.

Pao, C. V. (1978). Asymptotic behavior and nonexistence of global solutions for a class of nonlinear boundary value problems of parabolic type, *J. Math. Anal. Appl.* **65**, 616–637.

Payne, L. E. (1967). Isoperimetric inequalities and their applications, *SIAM Rev.* **9**, 453–488.

Payne, L. E. (1968). Bounds for the maximum stress in the Saint Venant torsion problem, *Indian J. Mech. Math.*, Special Issue, 51–59.

Payne, L. E. (1970). Some isoperimetric inequalities for harmonic functions, *SIAM J. Math. Anal.* **3**, 354–359.

Payne, L. E. (1976). Some remarks on maximum principles, *J. Analyse Math.* **30**, 421–433.

Payne, L. E. (1980a). Bounds for solutions of a class of quasilinear elliptic boundary value problems in terms of the torsion function, preprint,

Payne, L. E. (1980b). Isoperimetric bounds in a class of nonlinear elliptic problems, preprint,

Payne, L. E. and Philippin, G. A. (1977a). Some applications of the maximum principle in the problem of torsional creep, *SIAM J. Appl. Math.* **33**, 446–455.

Payne, L. E., and Philippin, G. A. (1977b). "Some remarks on the problems of elastic torsion and of torsional creep", Some Aspects of Mechanics of Continua, Part 1, Jadavpur Univ., 32–40, India.

Payne, L. E., and Philippin, G. A. (1979a). On some maximum principles involving harmonic functions and their derivatives, *SIAM J. Math. Anal.* **10**, 96–104.

Payne, L. E., and Philippin, G. A. (1979b). Some maximum principles for nonlinear elliptic equations in divergence form with applications to capillary surfaces and to surfaces of constant mean curvature, *Nonlinear Anal.* **3**, 193–211.

Payne, L. E., and Philippin, G. A. (1980). On maximum principles for a class of nonlinear second order elliptic equations, *J. Differential Equations* **37**, 39–48.

Payne, L. E., and Rayner, M. E. (1972). An isoperimetric inequality for the first eigenfunction in the fixed membrane problem, *Z. Angew. Math. Phys.* **23**, 13–15.

Payne, L. E., and Stakgold, I. (1972). "Nonlinear Problems in Nuclear Reactor Analysis." Proc. Conf. on Nonlinear Problems in Physical Sciences and Biology, Springer Lecture Notes in Math., Vol. **322**, pp. 298–307. Springer-Verlag, Berlin and New York.

Payne, L. E., and Weinberger, H. F. (1960). An optimal Poincaré inequality for convex domains, *Arch. Rational Mech. Anal.* **5**, 286–292.

Payne, L. E., Sperb, R. P., and Stakgold, I. (1977). On Hopf type maximum principles for convex domains, *Nonlinear Anal.* **1**, 547–559.

Philippin, G. (1978). Some remarks on the elastically supported membrane, *Z. Angew. Math. Phys.* **29**, 306–314.

Polya, G., and Szegö, G. (1951). "Isoperimetric Inequalities in Mathematical Physics." Princeton Univ. Press, Princeton, New Jersey.

Protter, M. H., and Weinberger, H. F. (1967). "Maximum Principles in Differential Equations." Prentice-Hall, Englewood Cliffs, New Jersey.

Protter, M. H., and Weinberger, H. F. (1973). A maximum principle and gradient bounds for linear elliptic equations *Indiana Univ. Math. J.* **23**, 239–249.

Rellich, F. (1940). Darstellung der Eigenwerte von $\Delta u + \lambda u = 0$ durch ein Randintegral, *Math. Z.* **46**, 635–636.

Sattinger, D. H. (1973). "Topics in Stability and Bifurcation Theory." Springer Lecture Notes in Math. Vol. 309. Springer-Verlag, Berlin and New York.

Saint Venant, B. De (1856). Mémoire sur la torsion des prismes, *Mém. Pres Divers Savants Acad. Sci.* **14**, 233–560.

Schaefer, P. W. (1977). On a maximum principle for a class of fourth-order semilinear elliptic equations, *Proc. Roy. Soc. Edinburgh Sect. A* **77**, 319–323.

Schaefer, P. W., and Sperb, R. P. (1976). Maximum principles for some functionals associated with the solution of elliptic boundary value problems, *Arch. Rational Mech. Anal.* **61**, 65–76.

Schaefer, P. W., and Sperb, R. P. (1977). Maximum principles and bounds in some inhomogeneous elliptic boundary value problem, *SIAM J. Math. Anal.* **8**, 871–878.

Serrin, J. B. (1969). The problems of Dirichlet for quasilinear elliptic differential equations with many independent variables, *Philos. Trans. Roy. Soc. London Ser. A* **264**, 413–496.

Simon, L., and Spruck, J. (1976). Existence and regularity of a capillary surface with prescribed contact angle, *Arch. Rational Mech. Anal.*, **61**, 19–34.

Sperb, R. P. (1972). Untere und obere Schranken für den tiefsten Eigenwert der elastisch gestützten Membran, *Z. Angew. Math. Phys.* **23**, 231–244.

Sperb, R. P. (1975). Extension of two theorems of Payne to some Nonlinear Dirichlet problem, *Z. Angew. Math. Phys.* **26**, 271–726.

Sperb, R. P. (1979a). Maximum principles and nonlinear elliptic problems, *J. Analyse Math.* **35**, 237–263.

Sperb, R. P. (1979b). Nonlinear diffusion-reaction problems with time-dependent diffusion coefficient, *Z. Angew. Math. Phys.* **30**, 663–675.

Sperb, R. P. (1980a). Growth estimates in reaction–diffusion problems, *Arch. Rational Mech. Anal.* **75**.

Sperb, R. P. (1980b). Isoperimetric inequalities in a boundary value problem on a Riemannian Manifold, *Z. Angew. Math. Phys.* **31**,740 – 753

Sperb, R. P., and Stakgold, I. (1979). Estimates for membranes of varying density, *Applicable Anal.* **8**, 301–318.

Steiner, J. (1882). "Gesammelte Werke," Vol. 2. pub. Berlin.

Ting, T. W. (1971). Elastic-plastic torsion of simply connected cylindrical bars, *Indiana Univ. Math. J.* **20**, 1047–1076.

Ural' Tseva, N. N. (1973). The solvability of the capillarity problem, *Book 19*, *Vestnik Leningrad. Univ. Math. Meh. Astronom.* **4**,

Walter, W. (1975). On existence and nonexistence in the large of solutions of parabolic differential equations with a nonlinear boundary conditions, *SIAM J. Math. Anal.* **6**, 85–90.

Watson, G. N. (1944). "A Treatise on the Theory of Bessel Functions," 2nd ed. Cambridge University Press, London and New York.

Yau, S. T. (1975). Isoperimetric constants and the first eigenvalue of a compact manifold, *Ann. Sci. Ećole Norm. Sup.* (4) **8**, 487–507.

INDEX

UNIV DUNELM